THE FOUNDATIONS OF
THE SCIENCE OF WAR

# The Foundations of the Science of War

By Colonel J. F. C. Fuller, D.S.O.

*Author of "Tanks in the Great War," "The Reformation of War," "Sir John Moore's System of Training," etc., etc.*

*The first Creature of God, in the workes of the Dayes, was the Light of the Sense; the last was the Light of Reason.*—FRANCIS BACON.

LONDON:
HUTCHINSON & CO. (*Publishers*), LTD.
PATERNOSTER ROW, E.C.

*Made and Printed in Great Britain by*
*The Camelot Press Limited,*
*Southampton.*

To
MY WIFE

# CONTENTS

## Preface

## DIAGRAMS

# PREFACE

The world is waking from its phantom dreams,
To make out that which is from that which seems.

—GERALD MASSEY.

## 1. THE ORIGINS OF THE BOOK

THE origins of this book may be of some interest, as the system outlined in it has been one of gradual growth, and, whatever value it may possess, it is the result of fifteen years' study and meditation.

In the autumn of 1911 I spent my leave in northern Germany, and returned to England convinced that a European war might break out at any moment. This realization stimulated my interest in military history, and to prepare myself for the inevitable and rapidly approaching struggle I turned to the *Field Service Regulations* (1909 edition) for assistance. On the first and second pages of Part I. I found the following:

> The fundamental principles of war are neither very numerous nor in themselves very abstruse, but the application of them is difficult, and cannot be made subject to rules. The correct application of principles to circumstances is the outcome of sound military knowledge, built up by study and practice until it has become an instinct.

This was excellent, but what were these fundamental principles? If they are neither numerous nor abstruse they must be few and simple, but not one was mentioned in the book, consequently it appeared to me that, unless I knew what they were, the *Field Service Regulations* was of little use. I determined, therefore, to discover these hidden truths.

I turned to the *Correspondence of Napoleon* and studied it closely, and during 1912 I had come to the conclusion that the principles which had guided Napoleon were as follows:

> . . . The principle of the Objective—the true objective being that point at which the enemy may be most decisively defeated; generally

this point is to be found along the line of least resistance. The principle of Mass—that is, concentration of strength and effort at the decisive point. The principle of the Offensive; the principles of Security, Surprise, and Movement (i.e. rapidly).[1]

I had now got six working principles, and, being satisfied with them, I was able to devote more time to Hall and Knight's elementary mathematics, the bugbear of the old Staff College examination, which I passed in the summer of 1913.

Whilst at the Staff College I applied my principles and found them a great help. Then came the war, and, in December 1915, I wrote an anonymous article for the *R.U.S.I. Journal* entitled "The Principles of War with Reference to the Campaigns of 1914–15." This article was published in February 1916, and to the former six principles I added two new ones—the principle of economy of force and the principle of co-operation. In the summer of 1917 General Kentish, who was then in command of the Commanding Officers' School in Aldershot, asked me to lecture on these principles, and I did so, and also on several other occasions. In March 1918 my lecture was published by him as a pamphlet.

So far these principles could only be looked upon as a pure hypothesis deduced from the campaigns of Napoleon and checked by the events of the Great War. In 1919 I was able to give them more thought, and I began to collect evidence in order to test them. This year a committee was assembled by the Army Council to rewrite the *Field Service Regulations*, and the chairman of this committee one day said to me: "I believe you have written something on the principles of war. May I have it?" I gave him a copy of the above-mentioned pamphlet. In 1920 the principles I had laid down were, in a slightly modified form, included in the new edition of the *Field Service Regulations*.

In July 1920 I wrote an article for the first number of *The Army Quarterly* entitled "The Foundations of the Science of War," in which my system was explained, and in 1922 I developed this system in chapter iii. of my book, *The Reformation of War*, which was published in February 1923. Between August 1922 and January 1923, being on half pay pending taking over an appointment at the Staff College, Camberley, I outlined and eventually wrote a series of some fifty lectures on "The Science of War" and "The Analysis of the Art of War." These lectures were given to the 1923 batch of Staff College Students, and were based on the following theory:

[1] See *Training Soldiers for War*, by the writer, p. 42. This little book was written in 1912 and 1913, and published in November 1914.

We start with man, and from man extract four elements:

| | | | |
|---|---|---|---|
| (i.) | Mental power .. | Mind | .. Control |
| (ii.) | Protective power | Protection | .. Stability |
| (iii.) | Offensive power | Weapons | .. Activity |
| (iv.) | Mobile power .. | Movement | .. Co-operation |

From these elements I evolved four elementary principles, namely:

(i.) From mind, the principle of the objective.
(ii.) From protection, the principle of security.
(iii.) From weapons, the principle of the offensive.
(iv.) And from movement, the principle of mobility.

I next postulated a law, which I called "The Law of the Conservation of Military Energy," and from it extracted four accentuating principles of war, namely:

(i.) The principle of surprise.
(ii.) The principle of economy of force.
(iii.) The principle of concentration of force.
(iv.) And the principle of co-operation.

Though these principles were of great assistance to me in working out problems in the physical sphere of war, it was difficult to apply them to mental and moral action. As regards mental action, I devised a co-efficient for each of them, and as regards moral action, from will, *moral*, and fear, I deduced three moral principles, namely:

(i.) The principle of determination.
(ii.) The principle of endurance.
(iii.) And the principle of demoralization.

In the autumn of 1923, having set these lectures together in book form, I submitted them to my friend, Captain B. H. Liddell Hart, and asked him to be unsparing in his criticism. This he certainly was, and his analysis of the MS. led to several prolonged discussions, particularly as regards the nature of the "threefold order" and the nomenclature of the principles of war. From his criticism I realized that the lectures were too complex, and that simplification was necessary. I consequently determined to rewrite the book, and if simplification has in any way been attained on the almost unexplored subject dealt with, I particularly wish to acknowledge my debt of gratitude to Captain Liddell Hart, and also to thank him for having read through and suggested amendments to the MS. of the book as it now appears.

I spent such spare time as I had in 1924 in reconsidering each step in my system, and it was not until January 1925 that I began to rewrite the book in its present form. A difficulty I unfortunately could not avoid was changing the names of some of my old principles, which, in the 1924 edition of vol. ii. of the *Field Service Regulations,* appear as follows:

(i.) Maintenance of the objective.
(ii.) Offensive action.
(iii.) Surprise.
(iv.) Concentration.
(v.) Economy of Force.
(vi.) Security.
(vii.) Mobility.
(viii.) Co-operation.

For the first I substituted the principle of direction, which is both more general and more accurate.

For economy of force I substituted the principle of distribution, and exalted economy of force to the position of the law of war.

I scrapped co-operation and introduced two new principles, those of endurance and determination, and left principles (ii.), (iii.), (iv.), (vi.), and (vii.) as they were.

I am of opinion that the whole system, though still far from perfect, has been greatly simplified by these changes. Though the principles have grown from eight to nine, they can, as I show in chapter xi., be reduced to three groups, namely, principles of control, resistance, and pressure, and finally to one law—the law of economy of force. Thus the system evolved from six principles in 1912 rose to eight in 1915, to, virtually, nineteen in 1923, and then descended to nine in 1925, with the added advantage that these nine can be merged into three, and these three into one law.

## 2. The Object of the Book

The book is what it is called, namely, a *foundation* of the science of war, or, at least, of *a* science of war, and, as I have spent over fifteen years in planning this foundation, I hope that military students will examine it, not only for its own worth, but in order to think of war scientifically, for until we do so we shall never become true artists of war.

I have stressed the scientific aspect of my subject, not that I am a trained scientist, for I am only an amateur, but because soldiers must realize what civil science means, and if, to-day, they spent half as much time in studying science, not forgetting a little philosophy, as they do in playing games, we ought to produce a very fine crop of generals.

To the scientist I have no doubt that my knowledge of science will prove limited, and possibly out of date, for, though I read a large number of scientific and philosophical works between the years 1898 and 1911, since 1912 I have found little time to continue this study; besides, I have seldom had the advantage of conversing with men of science.

In this book I have not attempted to apply my system historically; this I must leave for another volume; neither have I attempted, when dealing with the principles of war, to examine each principle in the same way. My examination may appear chaotic, but it is purposely so in order to accentuate the catholicism of these principles. A fault the critic will discover is repetition. Yet this again has been done on purpose, if only because Napoleon said: "There exists but one figure of speech for the crowd—repetition"; and Herbert Spencer said: "By iteration only can alien conceptions be forced upon reluctant minds."

Those who criticize this book must remember:

(i.) That the subject is all-embracing, and, consequently, must be incomplete.

(ii.) It is written in advance of the military thought of to-day.

(iii.) Many of the problems contained in it are very complex.

(iv.) And some of the terms I have used are vague, for scientific military terminology is sadly lacking in definition. Thus what is exactly meant by that semi-mystical word, *moral*?

To the civilian I think that this book may be of use, not only in studying war, but in studying any of the activities of life. As regards war, he must realize that everything is changing. We are faced by air warfare, and mechanical warfare on land, and submarine warfare at sea, and chemical warfare everywhere. What are the tendencies and values of these changes? This is not only a military question, but a national and an imperial question, for the defence forces exist for the empire, consequently every man and woman in the empire is *personally* concerned with their efficiency. To-day every other man (and still more so during war-time) is an amateur strategist and tactician; the House of Commons is full of such folk. No politician would be considered sane if he told a chemist or an astronomer what to do, but he considers it his right to tell the soldier, sailor, and airman what to do, and even how to do it; and if his words are not based on a true understanding of war they are based on a false understanding, for there can be no middle course.

Why this difference? It is because the soldier is ignorant

of his own profession. He will not use his brain save as an alchemist; consequently, the older he grows, the more his power of thought degenerates. As in "the stalactite caves of Carniola the blind salamander, Proteus, is found in great numbers, also blind assels, blind cyclopida, blind insects, and snails,"[1] so also in the fighting forces are to be found blind admirals, blind generals, and blind air marshals, because "any new set of conditions occurring to an animal which renders its food and safety easily attained seem to lead as a rule to degeneration. . . . Let the parasitic life once be secured, and away go legs, jaws, eyes, and ears; the active, highly-gifted crab, insect, or annelid, may become a mere sack, absorbing nourishment and laying eggs."[2] What a prospect for a Sandhurst cadet!

In this book I am attempting something new—at least, new since the days of Henry Lloyd and Robert Jackson; for, as far as I am aware, these are my only two fellow-countrymen who have attempted to reduce war to a science. In a small way I am trying to do for war what Copernicus did for astronomy, Newton for physics, and Darwin for natural history. My book, I believe, is the first in which a writer has attempted to apply the method of science to the study of war; for Lloyd, Jackson, Clausewitz, Jomini, and Foch did not do this. In a few years' time I hope that it will be superseded by many a better work, so that we all may begin to understand the nature of war, and thereby discover, not only how to prepare for war, but how to restrict its ravages; how to harness it, and possibly, also, how to transmute the destructive ferocity of the ape into the creative gentleness of the angel.

J. F. C. F.

Staff College,
Camberley,
*November 20th*, 1925.

[1] *The Open Court*, No. 105, p. 1803. [2] *Degeneration*, R. Lankester, p. 33.

# THE FOUNATIONS OF THE SCIENCE OF WAR

## CHAPTER I

### THE ALCHEMY OF WAR

Nothing is more terrible than active ignorance.—BOSSUET.

The art of war is like that of medicine, murderous and conjectural.—VOLTAIRE.

### 1. THE VALUE OF MILITARY HISTORY

THE history of war is a great romance, but as yet no true science of war has been written. For long the history of man and his perplexing ways were treated as a story, but in recent years the method of science has been applied to civil history, and to-day many historical works exist on the social, commercial, religious, and political evolution of nations. From these the student can discover, not only the sequence of past events, but their tendencies, and, above all, the probable direction of these tendencies in the future.

Though war is the oldest of the arts, no such method has as yet been applied to it. I will not say that attempts have not been made, for they have, but with little success; for most of the great writers on war lived before the advent of the present scientific age, and those who have written since have been obsessed by traditions. Guibert, in his *Essai Général de Tactique*, deplores "that whilst all other sciences are being perfected, the science of war remains in the cradle."[1] Lloyd, writing at about the same time, says: "It is universally agreed upon that no art or science is more difficult than that of war . . . yet those who embrace this profession take little or no pains to study it."[2] Robert Jackson, an English military surgeon, in 1804, sets out to examine the structure of war, "in order to inculcate useful truth" rather than "to furnish transient amusement."[3] His book still deserves study, and so does Lloyd's. Jomini is a great artist and geometrician of war, but little else, for he looks upon war mainly as "a

[1] *Œuvres Militaires de Guibert* (1803), vol. i., p. 97.
[2] *History of the late War in Germany* (1781), part ii., p. vi.
[3] *A Systematic View of the Formation, Discipline, and Economy of Armies* (1804), p. 12.

terrible and impassioned drama " [1]; yet, " I have seen," he says, " many generals—marshals, even—attain a certain degree of reputation by talking largely of principles which they conceived incorrectly in theory and could not apply at all."[2] Men, like General Ruchel, who, at the battle of Jena, thought " that he could save the army by giving the command to advance the right shoulder in order to form an oblique line."[3] Clausewitz, a military philosopher, never completed his great work, which is little more than a mass of notes, a cloud of flame and smoke; still, he writes of the art: " The conditions have been mistaken for the thing itself, the instrument for the hand."[4] At length we come to Foch, the most eminent soldier of our period, who, in 1903, sets himself this question: " Can war be taught? "[5] He believes that it can be taught, but only as an art based on theory. He quotes with approval the words of Dragomirov: " First of all, *science* and *theory* are two different things, for every art may and must be in possession of its own theory, but it would be preposterous to claim for it the name of a science. . . . Nobody will venture to-day to assert that there could be a *science of war.* It would be as absurd as a science of poetry, of painting, or of music."[6]

Surely it will not take more than a minute's thought to contradict this preposterous assertion. Poetry, painting, and music may be arts, but they are based on the sciences of language, of optics, and of acoustics. True, it is possible to be an artist without being a scientist, it is possible to theorize without knowing much, but this does not abrogate science, which, as I shall explain later on, is nothing else than true knowledge in place of haphazard knowledge, logical thinking in place of chaotic thinking, and, ultimately, truth itself in place of falsehood.[7]

[1] *The Art of War* (American edition, 1868), p. 360, also p. 344.
[2] *Ibid.*, p. 345.
[3] *Ibid.*, p. 57.
[4] *On War* (English edition, 1908), vol. ii., p. 130.
[5] *The Principles of War* (English edition, 1918), p. 1.
[6] *Ibid.*, p. 8.
[7] The confusion between the meanings of science and art in the head of the average soldier is most pronounced. They do not understand that " a science teaches us to know, an art to do " (Archbishop Thompson, *Laws of Thought*, p. 10); or that, as Professor Gore writes: " Every art is founded upon science; thus we have the science of electricity and the arts of electric lighting, electro-plating, etc., based upon it; the science of astronomy and the art of navigation dependent upon it; the laws of sound and the art of music. . . . There does not appear to be any real supernatural basis of any of the arts. Facts, laws, experience, and inference form the original source and foundation of all our knowledge, practice, and progress " (*The Scientific Basis of Morality*, p. 1). If an art is not based on science, then its foundations must be supernatural—that is, superrational. It is this alternative that such eminent soldiers as Marshal Foch have not considered.

Where are we to seek this theory of war which is unrelated to science? Foch answers: "History is the base," and then, approvingly, he quotes General de Peucker, who says: "The more an army is deficient in the experience of warfare, the more it behoves it to resort to the history of war, as a means of instruction and as a base for that instruction. . . . Although the history of war cannot replace acquired experience, it can nevertheless prepare for it. In peace-time it becomes the true means of learning war and of determining the fixed principles of the art of war."[1]

But, if we are disallowed a science of war, we can have no *true* history of war, only a "terrible and impassioned drama." On the battlefields we are artists of war, but we are seldom on the battlefields, for the greater part of our lives is spent in preparing for war in our lecture rooms, our studies, and on our training grounds. Here we are confronted by the history and mimicry of war. We do not want drama; we want truth. We require not merely a chronology of past events, but means of analysing their tendencies—means of dissecting the corpse of war, so that we may understand its mysterious machinery. To deny a science of war and then to theorize on war as an art is pure military alchemy, a process of reasoning which for thousands of years has blinded the soldier to the realities of war, and will continue to blind him until he creates a science of war upon which to base his art.

## The Reality of War

What, then, is the reality of war? For answer we must examine history. Wars come and they go; like flesh wounds, they ache whilst they last, and then, when they are healed, mankind forgets their smart. It is well that man should do so, for pain is an unpleasant sensation, so unpleasant that when we are wounded we pay large sums to those who can rapidly heal us.

In the past we have possessed innumerable witch-doctors of war, but few true surgeons, because we have possessed no science of war. The cauldron of war boils over; we are scalded; we shriek; some die; some recover; and then we lick our wounds and wait until it boils over again. Believe me, the history of war is an unbroken relation of these Medean performances.

If the student doubts my words, then let him read the history of the Crimean War, and he will find that the horror of its trenches, like some tragedy from the Grand Guignol, is, scene by scene,

[1] *The Principles of War*, p. 7.

replayed sixty years later in the swamps of Flanders. Let him read the account of the massacre of the Prussian Guard at St. Privat, in 1870. What does the Duke of Würtemberg say? He writes:

During the action at St. Marie aux Chênes, Prince Hohenlohe, commanding the Artillery of the Guard, had collected 84 guns opposite St. Privat, and cannonaded the French position with great effect, at first at 2,640 paces, and afterwards at 2,000 paces. *About five o'clock in the afternoon the Commander of the Guard considered the enemy to be sufficiently shaken for him to risk an assault across the open and gently ascending ground.* . . .

"The effect of the enemy's fire, even at a distance of more than 1,500 paces, was so murderous that, according to the accounts received, nearly 6,000 men fell in 10 minutes, and the advance had to be immediately discontinued."[1]

It is needless for me to remind the student that identical operations were carried out during the battles of Verdun and the Somme, forty-six years later. Forty-six years later! It is enough to make one weep!

Turn to the Russo-Japanese War: "At Shen-tan-pu the enemy made no less than five determined attacks against our entrenchment and its machine-gun, and were repulsed each time. The machine-gun did great execution, and we have heard—but this is not yet verified—that there were a thousand dead Russians left before it. At Li-ta-jen-tun the enemy could make no headway against our machine-guns, and was beaten back each time directly he tried to advance."[2]

Yet, in 1914, we had to learn the lesson of the machine-gun over again, and at what cost? We had to do so because war was looked upon as a dreadful drama, which required the most meagre of rehearsals for its preparation. "The truth is," writes Marshal Foch, "no study is possible on the battlefield; one does there simply what one *can* in order to apply what one *knows*. Therefore, in order to *do* even a little, one has already to *know* a great deal, and to know it well."[3] With this I full-heartedly agree; but I am of opinion that we shall never arrive at understanding war—that is, *knowing it well*—until we have a science of war which will reveal to us its reality, and not solely an art which must of necessity deal largely with its appearances.

[1] *The System of Attack of the Prussian Infantry* (English edition, 1871). Quoted in *A Précis of Modern Tactics* (1873), Major R. Home, p. 75.

[2] *Reports from British Officers attached to the Japanese and Russian Forces in the Field*, vol. ii., p. 56.

[3] *The Principles of War*, p. 6.

### 3. THE LACK OF THE SCIENTIFIC STUDY OF WAR

Though the scientific method has never as yet been applied to the history of war, truth always exists either openly or hidden; consequently its discovery is not so much a matter of knowing that effect B follows cause A, but *why* it follows. Long before James Watt watched the steam in his mother's kettle lift the lid, innumerable men had watched a similar phenomenon. Long before Sir Isaac Newton saw the apple fall, millions of human beings had shaken apple-trees to make apples fall. Yet these innumerable men and millions of human beings were not scientists, though Watt and Newton were, and, through discovering the laws of motion and of steam-pressure, they discovered truths, not necessarily absolute, but sufficiently general to enable thousands of artists (artificers of truth) to make use of them and apply them in a million ways.

Throughout the history of war, in spite of many famous artists, we look in vain for a military Newton or Watt. So much so that we see such eminent soldiers as Dragomirov and Foch affirming that war is solely an art and that there is no science of war. I think that I shall be able to prove that they are wrong, and that, because of this very ignorance of a science of war, the art of war has remained chaotic and alchemical.

If I am doubted, then again must I ask the student to turn to military history, and not merely examine one or two incidents as I have done, but read and re-read the campaigns of the great captains and study the operations of the great fools, for not only are these latter folk in the majority, but their art is immensely instructive. What will the student's verdict be? I imagine that it will agree with mine: namely, that we soldiers are mostly alchemists, and many of us little more than military sorcerers.

In the Great War of 1914–18 many of us witnessed curious happenings. Many of us partook of strategical black masses and tactical witches' sabbaths. Many of us sought the philosopher's stone and failed, and how ignominiously few of us as yet realize; for we, even to-day, possess no true test whereby to distinguish between the products of our ability and those of our incompetence. Be this as it may, do not let us despair of a little light, for as out of the twilight of the mediæval laboratory arose the great sciences of to-day, so out of this all but invincible ignorance may arise, if we so will it, a true science of war. It is for this reason that I have called this first chapter "The Alchemy of War," not because alchemy was utterly absurd, but because it was an art without a science. In alchemy what do we find? A false classification of real facts combined with inconsistent sequences—" that is, sequence not deduced by a rational method. So soon as science entered the field of alchemy with a true classification and a true method, alchemy was converted into chemistry and became an important branch of human knowledge."[1] So also with war; true facts have been examined, but their values have not been understood; and it is with these values that I shall deal in this book.

### 4. The Obsession of Traditions

It may be considered that I exaggerate the lack of war science in the past. Quite possibly I do; yet, outside the achievements of a handful of war geniuses, such as Alexander, Hannibal, Gustavus, and Napoleon, it is most difficult to arrive at the reasons of the military aims of the lesser captains. Either they set out to copy their masters, or else their battles were but matters of push of pikes, push of bullets, or push of shells. They were battles of imitation, or battles of brute force, and not battles springing from the foundations of a scientific knowledge of war. The main reason has been the obsession of traditions, for, as Sir Thomas Browne wrote in 1646, "the mortallest enemy unto knowledge and that which hath done the greatest execution upon truth hath been a peremptory adhesion unto authority and more especially the establishment of our belief upon the dictates of antiquity." In the opinion of this thinker, the universities, "though full of men," are oftentimes "empty of learning."

Marshal Saxe noted identical conditions in his day. Read his *Reveries*, and this is what he says:

> War is a science so obscure and imperfect that in general no rules of conduct can be given to it which are reducible to absolute certainties; custom and prejudice, confirmed by ignorance, are its sole foundation and support.

It would be difficult to write more sarcastically. Then he continues:

> Gustavus Adolphus invented a method which was followed by his scholars, and carried into execution with great success; but since his time there has been a gradual decline amongst us; which must be imputed to our having blindly adopted maxims, without any examination of the principles on which they were founded . . . from whence it appears that our present practice is nothing more than a passive compliance with received customs, the grounds of which we are absolute strangers to.[2]

[1] *The Grammar of Science*, Karl Pearson, p. 27.
[2] *Reveries upon the Art of War* (English edition, 1757), pp. iii., iv.

He suggested the reintroduction of armour, and he writes:

To say, then, that the enemy will adopt the same measures is to admit the goodness of them; nevertheless they will probably persist in their errors for some time, and submit to be repeatedly defeated for years, before they will be reconciled to such a change; so reluctant are all nations, whether it proceeds from self-love, laziness, or folly, to relinquish old customs: even good institutions make their progress but slowly amongst us, for we are grown so incorrigible in our prejudices that such, whose utility is confirmed by the whole world, are, notwithstanding, frequently rejected by us; and then, to vindicate our exceptions upon every such occasion, we only say, *'tis contrary to custom.*[1]

Such was the condition which prevailed before the Seven Years' War; and of the French and English generals during this war the French officer who translated General Lloyd's book into French writes:

One must obey these old fellows who, never having studied their profession, obsessed by an antiquated routine which they call experience, and taking advantage of a long existence which they consider a long life, set out to traduce, pull to pieces, and ridicule budding genius which they detest, because they are compelled to value it more than themselves.[2]

Such was the condition during the Seven Years' War. What, then, was the condition which followed it? To answer this question I will turn to another eminent soldier—Guibert—who, in 1769, published his *Essai Général de Tactique,* a book still worth studying. What does he say?

Of all the sciences which excite the imagination of men, the one concerning which most has been written, but about which the fewest books can be read with profit, is without possibility of contradiction—the science of war. . . . How happens it that no book has as yet appeared in which is laid down the principles of war? . . . I maintain that, from an instructional point of view, there scarcely exists a useful book on war.[3]

If these were the conditions which prevailed not only before but during and after the Seven Years' War, there can be little doubt, if we look back a few years, that they were the identical conditions which governed military thought prior to the Great War of 1914–18.[4] From the point of view of the science of war,

[1] *Ibid.*, pp. 46, 47.
[2] *Introduction à l'Histoire de la Guerre en Allemagne,* Général Lloyd (1784), p. xii.
[3] *Œuvres Militaires,* vol. i., pp. 129, 131, 135.
[4] See *The Science o War* Colonel Henderson chap. xiv.

progress had to all intents and purposes been stationary. The Germans were copying von Moltke; the French were trying to discover how to copy Napoleon; we—it is difficult to say what we were doing; we certainly were watching these copyists, and our thoughts were probably controlled more by French than by German military opinion.[1]

"The blind adoption of maxims." In these words of Marshal Saxe may be summed up nine-tenths of the art of war.

Because of Sedan, fought in 1870, the Germans, in the next war, were going to repeat Sedan on a scale tenfold greater. Because of Jena, fought in 1806, the French, in the next war, were going to repeat that magnificent manœuvre. Then, in 1914, before the war was six weeks old, these stupendous imitations dissolved into thin air.

The error here was not one of art—for the artist does a great deal of copying—but one of science, or, rather, one due to a lack of science.

Since 1806 and 1870 conditions had changed, and their values, which could easily have been ascertained by soldiers, were left undiagnosed, because armies were obsessed by traditions, and blindly adopted maxims.

### 5. The Foresight of Monsieur Bloch and Baron Jomini

It is always easy to be wise after an event, and, though this process must so often be resorted to, in the present instance I can quote from the written works of one man who, long before the outbreak of the Great War, because of his scientific training, was able to examine the nature of war scientifically. This man was not a soldier; he was a banker—Monsieur Bloch of Warsaw; and many soldiers thought him mad. In 1897 he published an immense work entitled *The War of the Future*; and in the introduction to the English translation of the first volume of this book we read:

> At first there will be increased slaughter—increased slaughter on so terrible a scale as to render it impossible to get troops to push the battle to a decisive issue. They will try to, thinking that they are fighting under the old conditions, and they will learn such a lesson that they will abandon the attempt for ever. Then, instead of a war

[1] Robert Jackson writes of the copyists of his day: "Hence, whatever relative excellence may actually exist between Prussian tactic and the tactic of other nations in their intrinsic merits, the professed copyist is still a copyist—not likely to attain a name in war, while he moves undeviatingly in the trammels of foreign institution. The principle of imitation expels the desire of novelty; yet novelty and change of form produce impression; and impression is the cause of success in war. Imitation discourages pride; but pride of mind is the essence of military virtue (*A Systematic View*, etc., p. 201).

fought out to the bitter end in a series of decisive battles, we shall have as a substitute a long period of continually increasing strain upon the resources of the combatants. The war, instead of being a hand-to-hand contest in which the combatants measure their physical and moral superiority, will become a kind of stalemate, in which, neither army being able to get at the other, both armies will be maintained in opposition to each other, threatening each other, but never being able to deliver a final and decisive attack. . . . That is the future of war—not fighting, but famine, not the slaying of men, but the bankruptcy of nations and the break-up of the whole social organization. . . . Everybody will be entrenched in the next war. It will be a great war of entrenchments. The spade will be as indispensable to a soldier as his rifle. . . . All wars will of necessity partake of the character of siege operations. . . . Your soldiers may fight as they please ; the ultimate decision is in the hands of *famine*. . . . Unless you have a supreme navy, it is not worth while having one at all, and a navy that is not supreme is only a hostage in the hands of the Power whose fleet is supreme.[1]

This forecast of coming events, made seventeen years before their arrival, is one of the most remarkable in the history of war, especially so as it was made by a pacifist. Monsieur Bloch was, however, so influenced by his own particular outlook, his maxim that war had become impossible through having become unremunerative, that he was content to consider his prediction as final. Had he been a thoughtful soldier, and had he possessed experience in the art of war, having analysed the nature of modern warfare, he would have arrived, I imagine, at the following conclusion : What was it that prohibited movement ? Fire-power ! What would protect the soldier against bullets ? Obviously, armour !

Here I will turn to another remarkable forecast made by one of the few really great military thinkers of the last century, namely Baron de Jomini. In his *Art of War*, written in 1836, this noted writer says :

The means of destruction are approaching perfection with frightful rapidity. The Congreve rockets—the effect and direction of which it is said the Austrians can now regulate—the shrapnel howitzers, which throw a stream of canister as far as the range of a bullet, the Perkins steam-guns[2]—which vomit forth as many balls as a battalion—will multiply the chances of destruction, as though the hecatombs of Eylau, Borodino, Leipsic, and Waterloo were not sufficient to decimate the European races.

[1] *Is War now Impossible ?* (English translation, 1899), pp. xvi.–lvi.
[2] This gun was invented by a Mr. Penn, and it was fired near the House of Commons to show the Duke of Wellington what it could do. I have as yet been unable to ascertain the date of this demonstration.

If Governments do not combine in a congress to proscribe these inventions of destruction, there will be no course left but to make the half of an army consist of cavalry with cuirasses, in order to capture with great rapidity these machines; and the infantry, even, will be obliged to resume its armour of the Middle Ages, without which a battalion will be destroyed before engaging the enemy.

We may then see again the famous men-at-arms all covered with armour, and horses also will require the same protection.[1]

The idea was excellent, but, at the time Jomini suggested it, it was quite impractical, for Jomini must have been fully aware that the main reason why armour had been discarded was that sufficiency of it could no longer be carried to protect the soldier effectively.

### 6. The Military Myopia Before the Great War

When Monsieur Bloch wrote his work on war the steam-engine had been brought to a high state of efficiency, and armoured traction-engines had already been built for service in Uganda, and, for tactical purposes, had been suggested as a means of destroying infantry in the Crimean and Franco-Prussian Wars. Further, the motor-car had just been born.

Now I maintain that had soldiers generally possessed the understanding to deduce the nature of the next war from existing facts—human nature as influenced by fire-power—as clearly as Monsieur Bloch had done, their answer to him would have been sought in the fulfilment of Jomini's prophecy.

Once it was realized that the unprotected infantryman could not face modern fire-power, then, knowing that half an inch of steel would stop a bullet, it needed but the most rudimentary common sense to see that armour should be reintroduced. As the horse and the man could not carry this armour, it would have to be carried for them. The only means of carrying it was some type of engine, and, as this engine would have to move off roads, it was clear that it would have to be furnished with caterpillar tracks.

Such machines were tested at Aldershot in 1907 and in 1908, but the military authorities could not see or foresee their use; for, in spite of the Russo-Japanese War, they were obsessed by the idea of a war of movement, and, in their opinion, these machines were too slow for a galloping horse!

What did the soldier see in the next war? A drama of glistening bayonets, a frenzied onrush of troops, a veritable Trojan

[1] *The Art of War*, pp. 48, 49. De Saxe and Henry Lloyd also recommended he reintro duction of armour.

contest. They laughed at Monsieur Bloch—the banker; and thus it was how France saw the approaching Armageddon:

> The war will be short and one of rapid movements, where manœuvre will play the predominating part; it will be a war of movement. The battle will be primarily a struggle between two infantries, where victory will rest with the large battalions; the army must be an army of personnel and not of materiel. The artillery will only be an accessory arm, and with only one task—to support the infantry attack. For this task it will only require a limited range, and its first quality must be its rapidity of fire, to admit of it engaging the manifold and transitory targets which the infantry will disclose to it. The obstacles which one will meet in the war of movement will be of little importance; field artillery will have sufficient power to attack them. In order to follow as closely as possible the infantry to be supported, the equipment must be light, handy, and easy to manœuvre. The necessity for heavy artillery will seldom make itself felt; at all events, it will be wise to have a few such batteries, but these batteries must remain relatively light in order to retain sufficient mobility, which precludes the employment of heavy calibres and powerful equipments. A battery of four 75 mm. guns develops absolute efficiency on a front of 200 metres; it is consequently unnecessary to superimpose the fire of several batteries. It will serve no useful purpose to encumber oneself with an over-numerous artillery, and it will suffice to calculate the numbers of batteries that should be allotted to the organization of formations on their normal front of attack.[1]

To-day, knowing what we do of the events of the Great War, it would be difficult to concoct, even as a joke, a more faulty appreciation, and when we compare it to the forecast of Monsieur Bloch, all we can do is to gasp!

What was the difficulty? It was that soldiers possessed no means of analysing facts; they saw things as cows see them, and they were unable to work scientifically. Had they been able to discover the true meaning—the truth—of facts, the rest of the problem would have all but solved itself.

### 7. The Military Myopia Since the Great War

The Great War cost us nearly one million dead, and it was concluded by a series of peace treaties which reek with future wars, yet, if we went to war to-day, we should do so with an equipment in several respects inferior to what we had in November 1918. What, then, have we learnt from this great upheaval? That war is such an unpleasant subject that the sooner we forget it the better; and, to make peace with its reason, the nation

[1] *L'Artillerie* (1923), Général Herr, pp. 4, 5.

chloroforms its intelligence by inhaling catchwords and meaningless maxims such as " the war to end all war " and " the abolition of war," when such absurdities can only end common sense.

Sometimes I almost despair of the future. During the Great War we saw tanks winning through, tanks just out of the cradle, imperfect machines which seldom could move more than four or five miles an hour. These machines, little better than standing targets, were faced by hundreds of guns. To-day tanks have attained a speed of over twenty miles an hour ; a British Division has but seventy-two field guns, and no infantry in the world will face a tank attack.

When, in our schemes and exercises, a battalion of tanks advances on a hostile division, that division, in spite of its seventy-two guns, is " dead meat " or " flying meat." Half the tanks may be put out of action, which is unlikely, nevertheless the remaining half will win the rubber. The reader may believe this or not, as he likes. All I can say is this : my opinion is based on the direct experience of at least a dozen tanks battles. In these battles I watched brave and efficiently trained troops—the German machine-gunners—literally melt away before tank attacks. In the future will infantry do better than they did, when faced, not by a machine crawling towards them at four miles an hour, but rushing on them at twenty-five?

Do we realize this ? If we do, then, for some reason or another, we are afraid to express our convictions, for, in vol. ii. of the 1924 edition of that useful book the *Field Service Regulations*, we read :

> Infantry is the arm which in the end wins battles. To enable it to do so the co-operation of the other arms is essential ; separate and independent action by the latter cannot defeat the enemy. . . . The rifle and bayonet are the infantryman's chief weapons. The battle can be won in the last resort only by means of these weapons. . . . The Lewis gun is a valuable auxiliary to the rifle.

This may be true in mountainous or thickly wooded regions, but it certainly is not true of fighting in open country. In the great artillery battles of the last war the infantry merely walked behind the barrage, and when the barrage stopped they stopped—they did not conquer ! In the great tank battles they merely walked behind the tanks, and when the tanks were knocked out, once again they stopped—they did not conquer ! To lay it down as an official doctrine that infantry is the supreme arm in all circumstances, and that the rifle and bayonet are still the supreme weapons of war, is in my humble opinion a dangerous over-statement.

I write this with a clear and definite purpose, namely, that, in spite of over four years of devastating warfare, few of us as yet have begun to realize the immense revolution which has taken place in the art of war. I believe that the main reason for this is that we possess no scientific method whereby to measure these changes.

In the past we have lulled ourselves to sleep on dogmas, and have been rudely awakened by realities which we have never troubled to foresee. Though we are soldiers, professing soldiership, most of us know no more about the science of war than a chimpanzee knows about the science of dynamics, though, as an artist, this brute excels in agility. It is for this reason that I intend to examine this subject; not to thrust my opinions down the throats of my readers, but to appeal to their imaginations, so that, by understanding the value of their art, war may be rendered more effective in the future, and, perhaps, less and less a dreadful and impassioned drama, and more and more a just and righteous force.

### 8. Opposition to Scientific Progress

There are two main causes for this military shortsightedness: the first is the worship of traditions, and the second is our incapacity to see world forces in their true relationship.

As regards the first, those of us who dare to disturb the dusty shibboleths of the past must be prepared, as history shows, to fight a somewhat sanguinary battle. It is not physical but moral courage we require, and that in abundance. The discovery of truth calls for brave men, for truth gives nothing to cowards. In the past all scientists have been attacked as heretics, and why? Because they were heretics. And not a few perished at the stake. When the stake had passed along its way, abuse and scorn replaced it, and to-day some of this former abuse appears so comic that I cannot refrain from quoting an instance.

Shortly after the Royal Society was founded a certain Mr. Crosse, vicar of Chew Magna in Somersetshire, declared it to be a conspiracy against both society and religion. "He regarded the use of the newly invented optic glasses as immoral, since they perverted the natural sight and made all things appear in an unnatural and, therefore, false light." He argued "that society at large would become demoralized by the use of spectacles; they would give one man an unfair advantage over his fellows, and every man an unfair advantage over every woman, who could not be expected, on æsthetic and intellectual grounds, to adopt the practice."[1]

[1] "On Some Aspects of the Scientific Method," F. Gotch. See *Lectures on the Method of Science*, p. 35.

Do we find such men as Mr. Crosse in the army? Yes—multitudes! He disliked spectacles; during the war I knew a major-general who was also an anti-optic fanatic; he disliked trench periscopes, and when, early in the war, a proposal was made to introduce them, he officially put down his objection on paper, and it read: "It is contrary to the traditions of the British officer to seek information from a position of security by means of a mechanical device"!

It is not the scientist but the alchemist who works like the natural philosophers mentioned in *Gulliver's Voyage to Laputa.* It may be remembered that one of these gentlemen wasted eight years of his life in attempting to extract sunbeams from cucumbers, in order to store them in hermetically sealed bottles and sell them during inclement summers. If for "inclement summers" we read "future wars," this method may equally well be attributed to the soldier.

As regards the second—our incapacity to see world forces in their true relationships—this fault has been not so much the soldier's as the civilian's. The civilian dislikes war, and he thinks that it can be killed by calling it by a bad name. Satan only exists when we believe in him. If we create a little hell and put war into it, it will take upon itself a hellish form, and, like a demon, it will annoy us. If, instead, war is looked upon as a world force, and we do not prejudice our views by calling it good or evil, we shall begin to understand it.

To-day it is pitiful to see the number of scientists, who pass as rational men, anathematizing war and urging men of science to have nothing to do with it. Their attitude is similar to that of the Church towards sorcery in the Middle Ages; and yet, when once persecution ceased, out of the witches' cauldrons bubbled the sciences of to-day.

To restrict the development of war by divorcing it from civil science is to maintain warfare in its present barbarous and alchemical form. To look upon war as a world force and attempt to utilize it more profitably is surely better. At one time, quite possibly, our ancestors were cannibals, yet hunger is not a vice, and even when a change over was made from eating vigorous young men and women to eating decrepit old people, this in itself was a distinct amelioration, which, in its turn, led to eating kids and lambs—yet hunger is still with us, and cannot be banished by a sigil or a decree. The moral needs no accentuation.

# CHAPTER II

## THE METHOD OF SCIENCE

Begin with observation, go on with experiments, and, supported by both, discover law and reason.—LEONARDO DA VINCI.

Must struggling souls remain content
With councils and decrees of Trent?
—LONGFELLOW.

### 1. AUTHORITY AND METHOD

LACK of science leads to chaos in art; I hope that I have made this clear. We must possess an art of war, and the truer this art is the more effective will be our actions. To teach an art demands a method of imparting knowledge, and, as an army should work like one man, method must be based on authority. Here, then, is our first difficulty, for authority to-day is largely based on unscientific foundations. The solution to this problem lies in simultaneously destroying and recreating authority. Our work may be compared to a serpent sloughing its skin; the old skin must not be torn off, but the process of forming the new skin must loosen the old and eventually detach it.

"Believe, and ask no questions," is the hub of a system which for many years I have fought against, yet the common mind asks for nothing better than to repose blindly in authority, and the common mind is not only to be found in the Higher Command, but in the rank and file as well; in fact, our whole military organization is obsessed by a military scholasticism which closely resembles the religious scholasticism of the Middle Ages.

To the scholastic, reason was but the handmaid of faith—an *ancilla fidei*—and surely in the present-day military world reason is still little more than a handmaid, for belief in the written word and unwritten tradition is still the master.

To me, the comparison between the mind of a twelfth-century monk and a twentieth-century soldier is so remarkable that it may be of some interest, for a moment, to consider the opinions of a few of those eminent and courageous men who battled against the chill, crystalline doctrines of the Middle Ages.

Bacon urged that authority must be disregarded, consequently he strenuously attacked the method of his age.

> Nor suffered living men to be misled
> By the vain shadows of the dead.[1]

Descartes, in his *Principia*, wrote: "The logic of the schools is only a dialectic which teaches the mode of expounding to others what we already know, or even of speaking much, without judgment of what we do not know."

Locke considered that scholasticism consisted in "empty verbalism and unverified assumption. . . . That every man see things as they are, and not merely through the eyes of others, was his greatest wish."[2] "Truth needs no recommendation," says Locke, "and error is not mended by it; in our enquiry after knowledge, it little concerns us what other men thought."[3]

To attack authority demands courage, but to replace the authority of assumption by that of reason demands a thinking man. The greatness of Bacon, of Descartes, and of Locke does not lie in their powers of destruction, but of construction. As Lewes says: "The special want of the age was a *method*, and these men furnished it." Therefore, as I consider that much of our present-day military theory savours of scholasticism, in order to follow in their footsteps I must also create as well as destroy, and if I only can create, destruction will follow as an inevitable consequent.

In this attempt to establish a method of studying war I realize full well that my machinery is imperfect; my reasoning may be faulty and my knowledge defective; I must ask, therefore, the student not to set authority lightly aside, but rather to rely on independent research in order that he may discover which is the more correct—authority or I. Research will lead to independence of thought, and this independence to an improvement of method—my own or someone else's. "It is not what the teacher does for the pupil, but what the pupil does for himself, that matters." This is the great lesson of Socrates, who suffered death because he was right and authority was wrong.

Before we cross swords with authority we must remember that an army is not a band of geniuses, but of ordinary normal men. Normal man, it should never be forgotten, is a product of fears and not of facts. He is a poor, receptive creature, obsessed by

[1] "Ode to the Royal Society," Cowley.
[2] *Scientific Method*, F. W. Westaway, p. 129.
[3] See *Ency. Brit.*, xiv., p. 756.

prejudices and fearful of novelty and innovation. As one writer puts it, we are surrounded by a "monstrous regiment of old men. . . . We prefer old judges, old lawyers, old politicians, old doctors, old generals, and when their functions involve any immediacy of cause and effect, and are not merely concerned with abstractions, we contentedly pay the price which the inelasticity of these ripe minds is sometimes apt to incur."[1]

All this and much else is due to normal men being in the majority. Their inclinations are static, and—I will repeat it again—an army is largely made up of such individuals, consequently power of judgment is never popular. This may be lamentable, but it is no use lamenting over it, for it is an irrefutable fact that the majority of mankind lives by imitation. Consequently the only common sense course open to us is to turn this limitation to our advantage by compelling men to imitate what scientific thinking has decided to be the most advantageous for a whole body of men, and not necessarily for each individual. In other words, we must discover and establish a common doctrine by a universal method. My object is, therefore, not to destroy authority, but to chasten it.

Method creates doctrine, and a common doctrine is the cement which holds an army together. Though mud is better than no cement, we want the best cement, and we shall never get it unless we can analyse war scientifically and discover its values. This, then, is the object of my method—to create a workable piece of mental machinery which will enable the student of war to sort out military values. Once these values are known, then can they be used like bricks to build whatever military operation is contemplated. My system, I believe, will enable the student to study the history of war scientifically, and to work out a plan of war scientifically, and create, not only a scientific method of discovery, but also a scientific method of instruction. Normal man *will not* think; thinking is purgatory to him; he will only imitate and repeat. Let us turn, therefore, these defects to our advantage; let us, through clear thinking and logical thinking, obtain so firm a mental grip on war that we can place before this unthinking creature a system which, when he imitates it, will reflect our intention and attain our goal. Let us look upon normal man as a piece of human machinery, a machine tool controlled by our brain. Let us devise so accurate a system, and let us present it to him in so simple a form, that without thinking, without perhaps knowing what we intend, he with his hands will accomplish what our brains have devised.

[1] *Instincts of the Herd in Peace and War*, W. Trotter, p. 87.

## 2. The Meaning of Science

Science aims at establishing the highest authority, and the man of science works by a well-defined method which is very different from the normal method made use of in the study of war, which, as I have pointed out, is similar, if not identical, to the method of the alchemists. I will now turn from this haphazard way of working to the scientific method, a system which, I think, will enable the soldier to evolve from the alchemy of war a science of war just as the science of chemistry was evolved from alchemy and kindred processes of work and thought. First, I will examine the meaning of science, for soldiers are so ignorant of the scientific method that I consider it wise to begin from the very beginning.

What is science? Science is co-ordinated knowledge, facts arranged according to their values, or, to put this definition still more briefly and to quote Thomas Huxley, science is "organized common sense," common sense being, in the opinion of this great thinker, "the rarest of all the senses."

"Wherever there is the slightest possibility for the human mind to *know*, there is a legitimate problem of science."[1] The result of this is that "There are no scientific subjects. The subject of science is the human universe; that is to say, everything that is, or has been, or may be related to man."[2] And, further: "Scientific thought is not an accompaniment or condition of human progress, but human progress itself."[3]

Bearing these facts in mind, it is beyond question that war, like all other human activities, may be examined scientifically, and it is in its examination, and not in what it may be in itself, that practical knowledge is to be sought, for it is a recognized fact that any branch of study "should be classed as a science, not in virtue of the nature of the things with which it is concerned, but rather in virtue of the *method* by which it pursues knowledge."[4]

In our study of war I maintain that our method has been a faulty one, and I maintain this, for in 1914 all armies were organically unprepared for war. These armies were not those which won or lost the war in 1918, and the difference between the tactical values of 1914 and 1918 is the measurement of the lack of scientific thought which characterized all armies before the outbreak of the war.

[1] *The Grammar of Science*, Karl Pearson, p. 17.

[2] "On the Aims and Instruments of Scientific Thought," W. K. Clifford, *Lectures and Essays*, vol. i., p. 141.

[3] *The Grammar of Science*, Karl Pearson, p. 45.

[4] "Psycho-Physical Method," W. McDougall, *Lectures on the Method of Science*, p. 113.

And how can science help us? What does it consist in? "It consists in strengthening, solidifying, and rendering conscious and coherent the ordinary processes of knowledge. The scientific man . . . claims to clear away fallacies, to bring into clear light the real principles by which all man's knowledge is acquired, and to use it."[1]

We discovered no principles, though we were always using the word. We saw many things, but we failed to classify and to correlate them; we did not discover the laws which govern military activities. Above all, we failed to criticize our opinions, and without criticism our ideas on war were not subjected to that refining process, the struggle for existence.

I realize full well that, whatever science of war we develop, it cannot be an exact science. War is primarily concerned with human acts; every fact is a new fact, nevertheless it is related to an old one of a somewhat similar type. In the physical sciences, facts are potentially independent of particular place and time, but in the study of war, as in the study of history, this is not so, since the greatest difficulty is to fix the human element. The spirit of man moves here and there and changes the complexion and value of things, yet the science of psychology is little by little discovering the hidden machinery of human actions. It is for this reason that I shall so frequently refer to the human element, and it is for this reason that the whole of my theory of war is based on man.

### 3. The Method of Science

To me, all that I have said is included in Huxley's definition of science, namely, "organized common sense." And common sense, what is this rare quality? Common sense is thought sentiment, or action adapted to circumstances, and circumstances are those innumerable conditions which surround us, some of which are stable and others in a state of perpetual flux. To work scientifically is to work in a common sense manner; and theories which are not based on common sense can be founded on nothing else than common nonsense—a condition which has been most marked throughout the history of war.

The scientific method of discovery is the common sense method, and "the aim of scientific thought . . . is to apply past experience to new circumstances." Surely this also is our aim in the study of war? What we want to know is the truth about the past, and then how we can apply this truth to the conditions

[1] "Scientific Method as Applied to History," T. B. Strong. *Lectures on the Method of Science*, p. 231.

which surround us and which will probably exist during the next war. "The scientific method is in itself meaningless," writes Professor Gotch; "it acquires merit through its aim, and is significant because of its purpose. Its form may, and indeed must, be plastic, varying with the conditions of man and of nature, but its end remains throughout the same—the revelation of truth about things." In brief, and to quote Virgil, the aim of the scientific method is expressed in the following line: "*Felix qui potuit rerum cognoscere causas.*"

To know the cause, or, rather, reason, this is to begin understanding truth. In war there are, however, so many things, that it would seem almost impossible to know where to begin. Once science was faced by a similar condition; but the scientist did not stand gaping at this difficulty; he began to organize knowledge, and so to form a base from which others could work.

In the study of war we are not as fortunate, for no one has shown us how to organize the facts of war. Hitherto we have, as artists, studied the technique of war, but "while technical thought or skill enables a man to deal with the same circumstances that he has met with before, scientific thought enables him to deal with different circumstances that he has never met with before."[1]

Here, then, is the supreme difference: If we can establish a scientific method of examining war, then frequently shall we be able to predict events—future events—from past events, and so extract the nature and requirements of the next war possibly years before it is fought.

The scientific method is, in my opinion, so important that I will quote what one writer says:

> The methods adopted by science are to obtain and record the facts in connection with any subject, to marshal and classify them in their proper relationship, and then to make a generalization which, in a brief but comprehensive formula, endeavours to account for the association between them and also the phenomena of their existence. As new facts are discovered they can be classified in their proper relationship, and interpreted easily and quickly with great economy of thought, while the properties of new or newly discovered substances and the results of newly observed phenomena may be predicted with a high degree of accuracy by applying to them the generalization—the theories—already formulated.
>
> But a fact which, seemingly, does not conform to the theory must be investigated further, or the theory must be discarded altogether in favour of a better generalization. The theory is the spirit of the fact, and must be in harmony with it.

[1] *Lectures and Essays*, W. K. Clifford, vol. i., p. 144.

It will be seen that science has no hard and fast line beyond which we must not trespass; the boundaries are constantly shifting with each new discovery, with more exact or more intimate investigation into phenomena, and theories are discarded unhesitatingly if the subsequent observations do not correspond with them."[1]

These, then, are the aspirations of the man of science: "He should deem no natural phenomenon too ignoble for investigation . . . he should grudge neither time nor labour in making and repeating observations and experiments . . . he should have the fear of error constantly before him, and . . . he should be unaffected by any considerations as to the immediate practical utility of his work. Free enquiry . . . conducted along these lines, guided throughout by man's most priceless possession, reason, and illuminated by his gift of imagination, has advanced scientific knowledge in the past, and will surely continue to advance it in the future."[2]

The whole of this method of science may be summarized in one word—" Experience," and it is with this word that I will now deal.

## 4. Observation, Reflection, and Decision

All knowledge is derived from experience, which includes the process of reasoning and imagination from the moment a sensation is received by the brain to the moment it is stored away in the memory. First there is sensation which at once gives rise to reflection[3]; so to say, the mind manipulates the sensation and the result is a decision, either conscious or automatic, that is uninfluenced by the will of the recipient. Those sensations which are perceived I will call observations. These are at once followed by an inference. For example, I hear a noise, and at

[1] "Scientific Management," H. Atkinson, *Engineering and Industrial Management*, vol. ii., No. 3, p. 71.

[2] "On some Aspects of the Scientific Method," F. Gotch, *Lectures on the Method of Science*, p. 58. Plato defines a philosopher as "one who gets inside things and discovers the nature of their reality, and contrasts him with those who are content with mere appearances and with ready-made opinion" (*Scientific Method*, F. W. Westaway, p. 24). "The philosopher," says Faraday, "should be a man willing to listen to every suggestion, but determined to judge for himself. He should not be biased by appearances; have no favourite hypotheses; be of no school; and in doctrine have no master. He should not be a respecter of persons, but of things. Truth should be his primary object. If to these qualities be added industry, he may indeed hope to walk within the veil of the temple of nature" (*Scientific Method*, F. W. Westaway, p. 49).

[3] Sensation awakens mental feeling; reflection gives rise to ideas. The difference was realized long ago in Plato's answer to Diogenes:

*Diogenes:* "I see a table and a cup, but I see no idea of a table or a cup."
*Plato:* "Because you see with your eyes and not with your reason."

once the nature of the sound heard suggests its cause. This is the beginning of reflection. I examine this cause, and it may appear to me unlikely, so I replace it by another, and ultimately arrive at a provisional decision. To prove this decision will demand a careful examination, not only of reasons, but of facts.

Experience may be said, therefore, to include three factors—observation, reflection, and their resultant, which is decision, the correctness of the sensation received being susceptible to proof by gaining contact with the cause of the sensation.

Accepting observation in its everyday sense, it is needless for me to say much, for its utility is self-evident. Some people are very observant, others see next to nothing; some only see small things, others only big, and most only see what others see, and what others see is very often not worth seeing.

The secret of observation does not so much lie in the quality of the thing observed, or in the quality of remembrance, as in the relationship of the thing to its surroundings at the moment of observation. To take a very simple example: a man on a cool day may walk twenty miles and show few signs of fatigue at the end of his journey; yet on a hot day he may show signs of collapse. The intelligent observer notices these two conditions, and, when he wishes to examine human movement, he remembers them as a relationship between human energy and temperature. The power of relating one thing to another is the foundation of reasoning.

Unless the student finds interest and is possessed with curiosity he will never observe. He will simply see things as a cow sees them, and, whatever grade he holds as a soldier, he will be but a military cow—every army is full of these beasts.

It is interest and curiosity which cause us to reflect, and if there is one word in the dictionary which is omnipotent it is the word WHY. Whatever I may say to the student, whatever he reads, whatever he thinks, he should ask himself the reason why. If he does not do so, however much he may strive to learn he will mentally be standing still. He must remember this: his brain is not a museum for the past or a lumber-room for the present; it is a laboratory for the future—a creative centre in which new discoveries are made and progress is fashioned.

Observation is the cause of reflection—that is, of reasoning—and it is only by reasoning that decisions are arrived at, and we must remember that a decision is something more than "Yes" or "No." If a judge were to omit taking evidence, and then say to the prisoner: "You are condemned," or "You are acquitted," he would cease to be a judge. When a general who has failed to reflect says to his subordinates: "You do this, or that," he ceases to be a general and becomes a dangerous maniac. Do not let us delude ourselves into believing that noises made with the mouth are necessarily decisions, for a decision is the offspring of reflection.

Science begins with observation, but observation must be methodical before it can be classed as scientific. "Every great advance of science opens our eyes to facts which we had failed before to observe and makes new demands on our powers of interpretation."[1]

## 5. Economy of Rational Thought

To cultivate the power of making sound decisions is no easy task. The biological process is that of trial and error, and this process results in adaption to enviroment and to evolution. The normal man works mainly by this process. He will watch others make a mistake a score of times, and then, in his turn, will make the same mistake. In fact, he learns next to nothing until he is made to suffer for his ignorance.

As man is the centre of the world of thought, and as thought governs action, and as it is visibly sound to economize the energy we expend, particularly during war-time, it stands to reason that we must begin by economizing thought. We must, in fact, establish an economical system of thinking before we can arrive at rapid and sound decisions.

> "The method of trial and error is a perfectly valid and legitimate one; it works. But it is costly and wasteful. It is cheaper to be wise, if we can, before the event than after it. Rational thought is the human improvement on the biological method of trial and error; a perfected, economical, immensely more effectual form of it. If one course of action proves successful and another fails, *there is a reason for it.* If sufficient knowledge had been available, if sufficient trouble had been taken, it would have been possible to know beforehand which was the rational and which the irrational course. The successful result is that to which efficient thought would have led had it been applied."[2]

Foresight, or the power of arriving at values before actions take form, is the highest form of judgment. When this power is inborn it is called genius—a subconscious realization of true values. Genius can be cultivated in a synthetic form, and, though this synthetic "substance" will not sparkle with the lustre of the natural product, it is a tremendous asset. Napoleon,

[1] *The Grammar of Science*, Karl Pearson, p. 45.
[2] *The Making of Humanity*, R. Briffault, p. 55.

one of the greatest war geniuses the world has ever seen, once said to Baron Roederer:

> "If I appear to be always ready to reply to everything, it is because before undertaking anything I have meditated for a long time—I have *foreseen* what might happen. It is not a spirit which suddenly reveals to me what I have to say or do in a circumstance unexpected by others; it is reflection, meditation."

Meditation was the one great secret of Napoleon's success, because meditation leads to rational thought, which within the sphere of rational things is always right. Rational thought knows no compromise or moderation, only the extreme view is right, because the ultimate extremity *is* truth. Thus, if I push a pencil off the table it will fall to the ground. This is a true fact; there is no compromise or moderation about it. It is facts of this kind we must strive to attain in our studies.

What is the main difficulty in attaining to this logical process of thinking? The difficulty is that we are slaves of the past; like monkeys, we are obsessed by imitation, we are for ever copying thoughts and actions without weighing their values or considering their results. The majority cannot learn, therefore aim to be one of the minority. Primitive man does not think at all unless by the direst necessity he is driven to do so; consequently do not hark backwards, look forwards. We must liberate our thoughts from customs, traditions, and shibboleths, and learn to think freely, not imitatively. When anything appeals to us or displeases us we must not accept it on its face value, but examine it, criticize it, and discover its meaning and inner worth. Remember that every student has much more to unlearn than to learn, and that he cannot learn freely until he has hoed the weeds of irrational thought out of his head.

### 6. The Machinery of Rational Thought

I will now turn to logic, or the machinery of rational thought, for, though I do not expect the student to study the numerous works written on the science of thinking, I consider that it is of importance that he should be able to recognize the leading methods.

When we think we are always inferring something—that is, making mental calculations. The first man who applied the scientific method to thought was Aristotle, who, in his *Analytics*, lays down three orders of inferences—analogical, inductive, and deductive. In the first order we infer from particular to particular, e.g. This thing has weight, so does that thing have weight.

If I say, however, that this thing has weight, so do all things have weight, I infer from particular to universal, and the process of thought is induction. If I reverse this, and say, As all things have weight, consequently this thing has weight, I infer from universal to particular, and the process is called deduction.[1]

Comte, the French positivist, compressed the essentials of all logic into the following maxim:

*"Induire pour deduire afin de construire."*

In other words, in order to construct rationally we must first work inductively and then deductively.

In modern times the inductive, or experimental, method was first studied by Francis Bacon,[2] who, warned by the failures of scholasticism,[3] propounded the following system:

(i.) Collect, observe, and tabulate phenomena.
(ii.) Note down all variations between them.
(iii.) By a process of exclusion the cause of any given phenomenon is discovered.

In brief, by means of the inductive method we attain to science by collecting facts, by sorting these into categories, by extracting their values, and on these values erecting theories. By putting these theories to universal tests, by degrees we extract laws which form our working principles, our weights and measures of war.

What Bacon attempted in the physical sphere Descartes attempted in the intellectual. He writes:

> Since we begin life as infants, and have contracted various judgments concerning sensible things before we possess the entire use of our reason, we are turned aside from the knowledge of truth by many prejudices; from which it does not appear that we can be any otherwise delivered, than if once in our life we make it our business to doubt of everything in which we discern the smallest suspicion of uncertainty.[4]

To Descartes the ultimate basis of knowledge was his own consciousness, and his fundamental axiom was "*Cogito ergo sum.*" I shall in my turn attempt to propound a somewhat similar (military) axiom in my next chapter.

In the examination of any problem Descartes lays down four rules of procedure[5]:

[1] "Induction is therefore the interpretation of facts, while deduction is the interpretation of sentences assumed to be true" (*Scientific Method*, F. W. Westaway, p. 171).
[2] See his *Novum Organum*.
[3] A system of philosophy which in the main subordinated thought to clerical interests. It was based on the works of Aristotle. Its exponents used deduction as their process.
[4] *Prin. of Phil.*, Descartes, i. 7.
[5] *Discourse of Method*, Descartes, part ii.

(i.) Never accept anything as true save what is evidently so.
(ii.) Separate everything into its component parts—analysis.
(iii.) Begin with the simplest components and work upwards to the more complex—synthesis.
(iv.) Make certain that nothing has been omitted.

Though the following was written of Bacon's system, it may equally well be applied to Descartes'. The lessons are:

> The duty of taking nothing upon trust which we can verify for ourselves; of rigidly examining our first principles; of being carefully on our guard against the various delusions arising from the peculiarities of human nature, from our various interests and pursuits, from the force of words, and from the disputes and traditions of the different schools of thought; the duty of forming our conclusions slowly and of constantly checking them by comparison with facts; of avoiding merely subtle and frivolous disputations; of confining our enquiries to questions of which the solution is within our power; and of subordinating all our investigations to the welfare of man and society.[1]

If in the mental sphere induction consists in tabulating, evaluing, and excluding, in the physical sphere it consists of examining, experimenting, and constructing. The greatest scientist of the last century—Charles Darwin—worked by this method. In 1837 he began his work—the discovery of the law of natural selection. He writes:

> By collecting all facts which bear in any way on the variation of animals and plants under domestication and nature, some light might perhaps be thrown on the whole subject. My first notebook was opened in July 1837. I worked on true Baconian principles, and, without any theory, collected facts on a wholesale scale, more especially with respect to domesticated productions, by printed enquiries, by conversation with skilful breeders and gardeners, and by extensive reading. When I see the list of books of all kinds which I read and abstracted, including whole series of Journals and Transactions, I am surprised at my own industry. I soon perceived that selection was the keystone of man's success in making useful races of animals and plants. But how selection could be applied to organisms living in a state of nature remained for some time a mystery to me.[2]

In 1838, due to a perusal of Malthus's *Essay on Population*, Darwin was inspired by the idea of a controlling law of selection. Between 1838 and 1842 he continued searching for facts, and criticized his hypothesis. In 1842 he put a brief abstract of his

[1] *Novum Organum*, Fowler, p. 129.
[2] *The Life and Letters of Charles Darwin*, vol. i., p. 83.

theory down on paper, but it was not until 1859 that he published his book, *The Origin of Species.*

In all, twenty-two years are spent in enquiry. First, facts are collected and examined; then a theory is propounded. This theory is subjected to prolonged criticism, and is eventually sufficiently proved to be classed as a law—the law of evolution.

I have quoted at some length the method applied by Darwin because I am convinced it is the model we soldiers should follow.

Induction is a simple and valuable process of reasoning, but, like all processes of thought, it has its limitations. In many subjects there exist too many alternatives for us to arrive at one universal, consequently the process of deduction from universals, either known or hypothetical, to particulars has to replace it. Professor Case takes, as an example, heat. In brief he says: " . . . by induction the nature of heat cannot be discovered. By the empirical method we know the phenomena of heat, and we know also that these are similar to the consequences of motion." In other words, we infer the nature of heat, not by induction, but by that kind of deduction which combines " phenomena " with " laws."[1]

The value of deduction is that:

(i.) It enables us to discover particulars inaccessible to the generalities of induction.

(ii.) It brings inductive facts under principles and so enables us to reach further than induction.

(iii.) It gives us greater power of discovering causation.[2]

In our study of war the deductive method will also help us, because we are confronted by innumerable facts the causes of which are generally unknown. Also it will help us, as it will enable us to make full use of our imagination—and this is essential in a science which is not an exact one, and which is interwoven so closely with the human element.

### 7. The Value of Imagination

Imagination is the telescope of our minds. It gives us distant glimpses of great things which can be handed over to the reason to analyse. Imagination must be controlled by method and founded on fact, yet frequently it enables us to discover causes and effects which, at the moment, are not rationally linked one with the other. Imagination works by hypothesis—that is, by assumption. Professor Tyndall tells us:

[1] " Scientific Method as a Mental Operation," T. Case, *Lectures on the Method of Science*, p. 12. [2] *Ibid.*, p. 13.

> Philosophers may be right in affirming that we cannot transcend experience, but we can, at all events, carry it a long way from its origin. . . . We are gifted with the power of imagination, and by this power we can lighten the darkness which surrounds the world of senses. Bounded and conditioned by co-operant reason, imagination becomes the mightiest instrument of the physical discoverer. . . . There is in the human intellect a power of expansion—I might almost call it a power of creation—which is brought into play by simple brooding over facts . . . the spirit brooding over chaos.[1]

Newton passed from terrestrial to celestial mechanics. " In the language of Tyndall, this ' passage from a falling apple to a falling moon ' was a stupendous leap of the imagination, for his enunciated law applies in conception to the universe, thus extending into boundless space and persisting through endless time."[2]

The hypothesis of the ether and the law of the persistence of force are stupendous assumptions, without which scientists could scarcely work. A hypothesis is not a vain speculation, for it must be based on facts and agree with their values. A hypothesis is a theory which binds facts together, a theory not only derived from the facts themselves, but also from their possible and probable conclusions. It is here that imagination based on reason comes to our assistance. Without some binding theory facts remain isolated and unfruitful; their contemplation should quicken the imagination ; for, as Sir Humphrey Davy once said : " It is only by forming theories, and then comparing them with facts, that we can hope to discover the true system of nature." Professor Jevons lays down three constituent conditions of a good hypothesis :

> (i.) A good hypothesis must allow of the application of deductive reasoning and the inference of consequences capable of comparison with the results of observation.
>
> (ii.) A good hypothesis must not conflict with any law of nature which we hold to be true.
>
> (iii.) In a good hypothesis, the consequences inferred must agree with facts of observation.[3]

In brief, the method of science is based on analysis, synthesis, and hypothesis, the one necessarily involving the other. We first observe ; next we build up a hypothesis on the facts of our observation ; then we deduce the consequences of our hypothesis

[1] *Fragments of Science*, Tyndall, vol. ii.

[2] " On some Aspects of the Scientific Method," F. Gotch, *Lectures on the Method of Science*, p. 54.

[3] *Principles of Science*, Jevons, p. 510.

and test these consequences by an analysis of phenomena ; lastly we verify our results, and if no exception can be found to them we call them a law.

Without imagination the man of science lacks mental vision. " All great scientists," writes Professor Pearson, " have, in a certain sense, been great artists ; the man with no imagination may collect facts, but he cannot make great discoveries." Imagination leads to " the discovery of some single statement ; some brief formula from which the whole group of facts is seen to flow is the work, not of a mere cataloguer, but of the man endowed with creative imagination. . . . The discovery of law is therefore the peculiar function of creative imagination. . . . Hundreds of men have allowed their imagination to solve the universe, but men who have contributed to our real understanding of natural phenomena have been those who were unstinting in their application of criticism to the products of their imaginations. It is such criticism which is the essence of the scientific use of the imagination, which is, indeed, the very life-blood of science."[1]

If criticism is the life-blood of science, then of all the weapons in our mental armoury it is the most potent in our study of war. Hitherto (and still to-day) in our army criticism has been looked upon as a breach of discipline. To criticize the actions of a noted general, especially if he be alive, is considered derogatory to military etiquette, and the result is that without criticism there can be little or no progress, and without criticism strategy and tactics must remain alchemical arts. The man who cannot support criticism is a man who dares not look into the eyes of Truth. What did Cousin say ? He said :

" LA CRITIQUE EST LA VIE DE LA SCIENCE ! "

Let the student remember these eight words, and make them his guiding star in his study of war ; and, if he be wise, let him remember also the words of a still greater man—Galileo :

" WHO IS WILLING TO SET LIMITS TO THE HUMAN INTELLECT ? "

The man who does petrifies his brain.

[1] *The Grammar of Science*, Karl Pearson, pp. 37, 38.

# CHAPTER III

## THE THREEFOLD ORDER

The general order, since the world began,
Is kept in nature, and is kept in man.—POPE.

There is but one temple in the Universe,
and that is the Body of Man.—NOVALIS.

### 1. THE FOUNDATIONS OF KNOWLEDGE

IN the first chapter of this book I showed, and I think beyond dispute, that it was not so much the lack of knowledge, but of method in its examination, which has rendered the study of war so chaotic. Now if, before applying the method I have summarized in the last chapter, I can establish a foundation so universal that it may be considered axiomatic to knowledge in all its forms, then, not only shall I be able to work from a solid base, but I shall be able to bring the study of war into the closest relationship with the study of all other subjects. If this foundation is so layed out—as I believe it to be—that from its outline can be perceived the form and proportions of its eventual superstructure, then I shall possess a guide towards design and a key-plan to work by.

In the examination of these foundations I must, perforce, enter into a little elementary philosophy, since philosophy embraces universals, but, in so doing, I intend to establish my base in as simple a manner as I can, since my object is to assist military students and not philosophers.

The first question which confronts us is: What is the ultimate source of knowledge? My answer is: For a moment let us look around and think, and we shall soon realize that the world as it appears to us is unceasingly surging from rest to activity, and from activity sinking back into a state of restfulness. We sense a continuous, never-ending pulsation. What, then, is its rhythm?

Complete inertia and absolute activity are unthinkable qualities, and whether the world is evolved from a single source or from two or more separate sources does not concern us here, since thought cannot penetrate beyond duality. For a thing to exist within the limits of our consciousness, which is the relationship between the ego and the non-ego, it must possess two opposite poles or extremities. Both these poles are in themselves incomprehensible, for the only factors which the mind can grasp are the relationships between their differences.

The nature of all knowledge is, therefore, relative; that is to say, it is only the record, or a reflection, of the interplay between the differences of these two poles, and this relationship is a dual one. Thus, if I am represented by A and the universe by B, the relationship between myself and the universe is subjectively $+$ AB, and objectively $-$ BA. A does not exist apart from B, neither does B exist apart from A, nor can their relationships exist apart from either, since all three exist as a trinity in unity, and it is this triunity which enables us to know. Knowledge is, in fact, based on the universal inference of a threefold order —this is my *cogito ergo sum*.

Having established this hypothesis, I will now attempt, not to prove it, as it must always remain an assumption, but to render it more tangible.

If I look upon the universe as space of three dimensions, then this space manifests to my mind in terms of time and force; time including the subjective relationships of mind and space, and force—the objective relationships. Time may be divided into past, present, and future; and force into energy, motion, and mass. We only know the past through the present, and can only speculate as regards the future from the present; and all our subjective knowledge in time is ultimately based on objective motion, or the relationship at any given moment between energy and mass.[1]

Because of the mind, about which we practically know nothing, we become conscious of the present and of motion, and through the present of the past, and, to a lesser extent, of the future; and through motion of mass, and also, I think, to a lesser extent, of energy. When some event happens again and again, we infer that it will happen yet again, and this inference, when we have discovered the reason why it happens, we call knowledge; or, if we are not certain of the reason, we assume that it will happen again, and this assumption we call belief. The relationship between knowledge and belief I will call faith, and if knowledge is A and belief is B, then faith may be either A—B or B—A. Whatever metaphysics may demand, what scientific faith requires

[1] When we think of time as eternity—that is, timeless time—or space as vacuum—that is, space devoid of matter—we are only thinking in abstractions rendered possible by what I will call common sense time and space, that is, time which to the human mind is never fixed and space which is never empty.

is that A should be as great and B as small as possible—yet there must always be some B.

I will now set down my argument in graphic form:

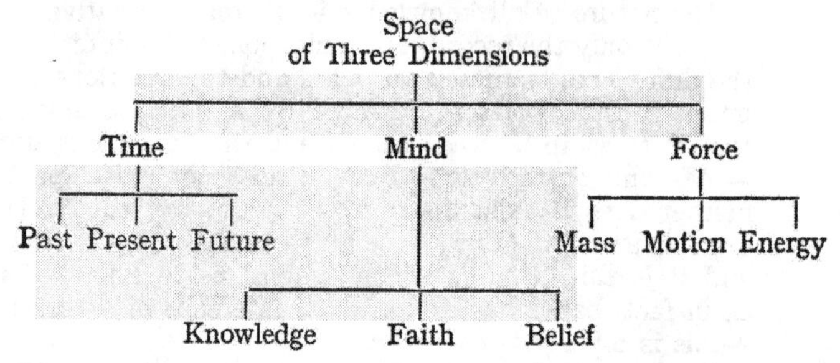


Reason ultimately is based on an assumption; therefore, strictly speaking, all knowledge is assumed. Common sense accepts this situation, nevertheless the common sense thinker differentiates between assumptions; he knows that he knows more about the past than the future, and about mass than about energy, and that the first two are realizable in what he calls the present, and the second two in what he calls motion. In the past and in mass he finds something concentrated and tangible, and in the future and in energy, something distributed and less easily grasped.

If we now turn to mind, it is scarcely necessary to explain that knowledge is mainly the product of analysis, and assumption of hypothesis, and that faith is the synthesis resulting from the relationship between these two. Faith is our directing force; I have faith in my knowledge or my belief, and thus faith is my guide through life.

I hope that this brief examination has made my meaning of the threefold order clear, an order which flows like an electric current between the poles of inertia and activity, and which is measured in terms of change. The human mind deals with change—changes of motion in an ever-changing present, and the terminals in themselves remain unknowable. The world as it appears to us is, therefore, but a reflection of the world as it is in itself, and, as absolute knowledge of the world is not vouchsafed to our reason,[1] consequently all our knowledge is but relatively true, as true when compared to the Absolute as my

[1] William James writes: "The 'absolutely true' meaning what no further experience will ever alter is the ideal vanishing-point, towards which we imagine that all temporary truths will some day converge. It runs on all fours with the perfectly wise man and with absolutely complete experience, and, if these ideals are ever realized, they will all be realized together" (*Pragmatism*, p. 223).

reflection in a mirror is true when compared to myself. In the mirror my left side becomes my right; in Reality it is possible that my inside (centre) becomes my outside (circumference), and that things can be known centrally and not merely circumferencially.

To pursue this question further would be to digress, for the subject before me is common sense knowledge and not metaphysics. We live in a three-dimensional world, and our knowledge is based on a threefold order. There may ultimately be an absolute plus and an absolute minus, a complete state of activity and of inertia. We cannot, however, grasp these states, but only the changes in the current which flows between them. We start from some conventional zero, and, by working upwards or downwards, we give plus and minus quantities a measurable meaning—that is, a relationship within our minds.

This threefold order surrounds us at every turn. Not only do we live in a three-dimensional world, but we think three-dimensionally and our thoughts reflect a threefold order. We sense ourselves as mind, body, and soul, and the world as force moving through space. We talk of God, Nature, and man; all our religious ideas are ultimately based on a trinity, as are those of all but the crudest of cults. We see Nature as earth, water, and air, and mankind as men, women, and children. We are surrounded by solids, liquids, and gasses, and by birth, life, and death. We live in a perpetual twilight, that infusion of light and darkness which in themselves are, to our minds, zero—that is, they are incomprehensible. This threefold order I believe to be the key to the understanding of all things; it is my postulate.

This threefold order forms the axle-pin of my system, which, I hope, will enable the items of war to be more readily evalued than heretofore. In this system, in place of making use of the term inertia, I shall generally talk of stability. To me stability denotes resistance, and activity opposition to resistance—that is, pressure. The changes, or movements, between these two are the resultant of their co-operation. Thus, if I wish to break a stick, I place it across my knee and pull it towards me. The stick is possessed of stability, my muscles of activity, and the relation between these two—the tension, the strain, and the ultimate snapping of the stick—is the movement generated by the co-operation between the resistance of the stick and the pressure exerted by my muscles.

Whether this threefold order is a universal law I am not prepared to say, but as it forms the norm of my entire system, if it is overlooked, the system itself will be difficult to understand. I will now turn to the brain of man—the storehouse of knowledge.

2. The Storehouse of Knowledge

Knowledge is a brain culture and not a world culture, and the brain, like a heat-engine, cannot work without a relationship between two differences. I have already stated that thought cannot penetrate beyond duality, consequently without duality it is not possible to conceive of reason, which is the relationship between mind and the outer world, or between mind and mind, or between the thoughts within the mind which the outer world has forced the mind to store. These relationships constitute knowledge, which is piled up in the mind in the form of accumulated mental work or mental energy, the economical expenditure of which is the most important problem in war, as it is in all the other activities of life.

The actual storehouse of knowledge is called memory—conscious or subconscious. Then, when the mind mobilizes its thoughts, the threefold order takes form, and the thread of plus quantities is woven through the woof of minus quantities; thus are ideas formed and decisions arrived at.

This storehouse is filled by study, by experience, and by information. One of our main sources of study is history, in which is collected the past experience of others. To read history is not sufficient, for history is full of assumptions and errors; therefore, unless we can deduce the reasons for these assumptions and evalue the events recorded, and apply these reasons and values to our present and future problems, our reading will be of little use. In place of reading history we must study it—that is, we must think over the relationships between the items which go to build it up, and from observation and reflection arrive at a decision regarding them. Locke, very truly, says: "Reading furnishes the mind only with materials of knowledge; it is thinking makes what we read ours. The memory may be stored, but the judgment is little better, and the stock of knowledge not increased, by being able to repeat what others have said."[1] We must work on a system! To-day we have no system, and it is my intention to create one.

I will now examine experiences which simultaneously possess great values and dangers. Their main values are those of a mental rather than of a physical character, and especially so in war. Thus, it is not so important to realize the physical results of certain actions as it is to know the state of mind which was induced by them during their execution. The reason for this is a simple one, and I will explain it by an example.

There is no difficulty in understanding the protective value

[1] *Conduct of the Understanding* section xx.

of an artillery barrage, but, to those who have never experienced walking behind a "wall" of bursting shells, it is next to impossible to realize what it morally "feels" like. Again, anyone can picture to himself the physical effect of machine-gun fire; but in peace-time it is not practicable to experiment on human nerves by actually firing at a human target. We thus find that, to those who work alchemically, experience is generally a danger rather than a blessing. Whilst *matériel* is always changing, nerves remain constant, or nearly so, consequently the most permanent lessons the experiences of war should teach us are those of a moral nature; yet in peace-time these are the more rapidly forgotten, since we possess no system which will balance the mind.

The dangers of war experiences are to be sought in their novelty and vividness; they are apt to obsess an unbalanced mind and leave it spellbound. We see something accomplished which leads to success or failure, and we judge of it by results, with little reference to the circumstances of the moment, which frequently are unknown to us.

In war nothing is more dangerous than jumping to conclusions on isolated actions, or of basing a theory on a single success or failure. What proves a success in one set of conditions may well prove the greatest of failures if these conditions be slightly shuffled. This fact history bears record to again and again, so frequently that it may with truth be said that a common cause of disaster is the copying of methods which in the past have proved themselves successful. Again we arrive at the necessity for some system which will enable us to correct our thoughts and discover the true meaning of events and experiences.

Lastly, as to information, which is the contact of mind and mind, and not of mind with the other world, or of thoughts within the minds. Here we are presented with knowledge in the second degree. In war we have largely to rely on information, consequently if the two minds be differently trained, as they usually are, and if they are collecting knowledge on a different system, or, what is more often the case, on no system at all, values will become mixed, and time will be wasted in untying these mental knots. To take a simile, each brain is constructed to resemble a photographic camera; but, unless each camera is in focus, the negatives will not be similar. That this focus seldom exists in the untrained mind is readily proved by the proverbial unreliability of eye-witnesses, and the history of war is largely built up on their evidence. Yet I believe that, if observation is systematized, reliability can be established; and, if reliability is attainable, reflection can be simplified and truer decisions arrived at. Here again we require a system, and one which will not only train men to see the things they are required to see, but to think of them from a common basis.

I will assume that potential knowledge in its totality is unlimited, or so vast that at present man's brain has only rendered a fraction of it conscious. By inference we assume that a few years hence our knowledge will be greater than it is to-day. Progress means stepping forward, therefore past knowledge is our base of action from which with some assurance we can attack our ignorance and transform it into knowledge. Thus our present knowledge becomes our means of action as well as our stable base, and if this knowledge is systematized so that we can correctly analyse past knowledge, by turning our minds forward and by making use of this same process, we are able in many cases to predict the nature of future discovery, and so advance in our knowledge more rapidly than if we leave discovery to chance.

Given the threefold order as a guide, the question now arises: Is there any prototype which will provide us with a key-plan to work to? I believe there is.

### 3. The Architypal Organization

To me the one great measuring-rod is the body of man, for, with Protagoras, I believe that: "Man is the measure of all things." All the knowledge we gain is through our minds, toned by our souls and expressed by our bodies. All the change we effect and the inventions we introduce are made to assist and enlarge our natural abilities. The world which man knows is of his own creating. Everything he thinks and does is measured out in proportion to his natural powers; in fact, the world *he knows* is a radiation of himself. The illusion is that he does not realize this, and, when he beholds the world he has created, he thinks of it as something apart from himself, and then he attempts to organize it on lines which do not reflect his measurements. Nevertheless, in spite of this inverseness, his world and his work are always tending to approximate in organization to his own body, which is the most wonderful and perfect machine devised, a fitting temple for his intelligence to inhabit.

Though human inventions and discoveries astonish us daily, the body of man still remains the most wonderful piece of automatic machinery in the world, and for many centuries yet to come will man's mind be concerned in examining its works. The most mysterious of events which daily takes place is the procreation of life, and the workshop of life is so marvellously organized that to overlook it as a model is to me all but a blasphemy.

Whatever we are asked to organize, we should think in terms of the human body, for as the world is a reflection of Something on the mind, so should all human organizations reflect the threefold order in man.

### 4. The Threefold Organization of Man

I will now take man as my model and examine him in a common sense way, a way which can be employed by anyone, even if his knowledge of physiology be of the slightest.

First I see man as an object—a *body*; then I find that this body is not inert, but conscious; it possesses a *brain*; and then somewhere in man lives his *soul*, or ego, which, by endowing him with character, differentiates him from his fellows.

Once again are we confronted by the threefold order, and, bearing this in mind, I will now turn to the human body and examine it. What do I find? That it is based on a threefold organization: it possesses structure, and powers of control and of maintenance. Thus:

(i.) *Structure.* The body, as we see it, is a compound of bones, ligaments, and muscles. The bones give stability to the whole organization; they keep it erect and in shape. Without bones man would be but a human jelly-fish. The ligaments bind bone to bone and muscle to bone, and enable the muscles to work or co-operate with the bones. The muscles give flexibility to the whole organization, yet their activity would be negligible if they were deprived of the bones upon which their actions are based.

(ii.) *Control.* The body is controlled by the brain, one part of which automatically governs the internal organs, and another part consciously regulates the limbs and external organs—eyes, ears, etc. Its functions largely depend on the information gathered up by the senses, and conveyed to it by the nerves, and also by means of the nerves it regulates the movements of the body.

(iii.) *Maintenance.* The body is maintained by the internal organs, of which the power-house is the stomach. Here energy is distributed to the body by means of the blood; and the tissues are repaired, and the waste products collected by various organs and ejected.

If, now, from the apex of this organization we look downwards, we shall see that each main organic division possesses power of action which is expressed by co-operating with a stable base and working from it. Thus:

(i.) In the structure of the body:
(*a*) The skeleton is the stable base.
(*b*) The muscles possess power of action.
(*c*) And the ligaments enable activity to become manifest by linking muscles and bone in close co-operation.

(ii.) In the control of the body:
(*a*) The senses form the stable base, or source of information.
(*b*) The brain possesses power of action.
(*c*) And the nerves enable activity to become manifest by linking brain to muscle in close co-operation.

(iii.) In the maintenance of the body:
(*a*) The stomach forms the stable base, or source of supply.
(*b*) The repair and evacuative organs possess power of action.
(*c*) And the blood enables this activity to become manifest by linking the stomach to all parts of the body in close co-operation.

I realize that these deductions are in nature very general, but, if they are moderately correct, we may, I think, from the body of man abstract three qualities, or elements, namely:

(i.) The element of stability (the negative element).
(ii.) The element of activity (the positive element).
(iii.) And the element of co-operation (the relative element).

These three, when correlated, build up the human organism. The aim of every living thing is to continue to live, and this object is striven after through the closest possible interplay between the above three elements. Power to move cannot become manifest unless it is based on a stable foundation, or frame, and linked to this frame by the element of co-operation. Granted this link, movement takes place when the stable and active elements are in co-operation; man is, in fact, a human engine which can move from place to place or stand still at will.

### 5. The Threefold Nature of Man

The brain of man is the controlling organ of his anatomy, yet it is not a free agent, for its control is accelerated and retarded by what I have called the soul of man. The brain of man is continually being bombarded by impressions, and the soul of man is the focal point of this bombardment. Each of these impressions changes man, and not only his mind, but his character. Though I cannot here enter, even superficially, into the values of normal psychology, I consider it of importance that the threefold nature of man should be realized, since wars, like all other human activities, are matters of men and the wills of men in harmony or in opposition.

Man is a compound of soul, mind, and body, three modes of force which must be expended, controlled, and maintained in war. I will now briefly examine these forces.

(i.) *The Soul of Man.* Every living organism, however primitive it may be, possesses feeling, or power of becoming aware of itself as an existence apart from its surroundings. When an outer object is brought into contact with it, a feeling or sense-impression is produced, and a sensation results which, according to its quality, the pleasure or pain it stimulates, becomes a desirable or undesirable sentiment. Should this sentiment become fixed through repetition, it is called habit; if through hereditary action, instinct. The strongest instinct evolved by natural selection is the instinct of self-preservation.

(ii.) *The Mind of Man.* Each sense-impression leaves on the substance of the feeling a trace, or mark, which is retained by a quality of the mind known as the memory. The interplay between memories results in thought and between the ideas in imagination. The interplay itself is known as the understanding, or the power of tracing causation; and the faculty which renders this interplay possible is the reason. Reason is the faculty of thinking; and when thoughts are fixed in one direction by a conscious impulse the result is will—the motor-force of the organism which produces it.

(iii.) *The Body of Man.* Will, once set in motion, is directed by purpose, and leads to a definite act, which is the material or outer effect of the psychological or inner cause. The immediate agent of this act is the body, and particularly the movements engendered by the muscles. These movements may be classified under two headings: voluntary, or conscious movements, and involuntary, or subconscious. Subconscious movements are of two kinds: instinctive movements, such as that of a newly born child seeking its mother's breast, and acquired movements, such as a man guarding himself in fencing.

From this brief summary it will be seen that man is possessed of three spheres of force; his mind works in the mental sphere, and his soul and muscles in the moral and physical spheres respectively. This may seem a very obvious discovery, and one of no particular importance. It is, however, one of the greatest importance, not that I have discovered anything—I have not—but what is of importance is that later on I am going to apply this discovery to war. Whenever I think of force, I am going to think of it in terms of these three spheres of force, which are a trinity and, consequently, can never be separated.

If this brief examination of the threefold nature of man is accepted as being correct, then it follows that, because man, in common with all other animals, possesses a quality called feeling, which is susceptible to sensation, sensations become the source of all knowledge and of all moral characteristics. In the mental sphere a sensation takes the form of thought, which is a reflection of the object sensed. In the moral sphere it is the quality of each sensation which endures, and not its form. Whilst thought is controlled by our power of reasoning, which may lead to true or erroneous decisions, sensations are moulded by our power of sentiment into pleasurable and painful qualities; normally the first are beneficial and the second harmful to the health of man. A mental decision leads to a physical action, actions being the concrete and tangible manifestations of our thoughts. Actions may be constructive or destructive, the controlling power being our muscles. We thus obtain three spheres of force, which diagramatically may be shown as follows:

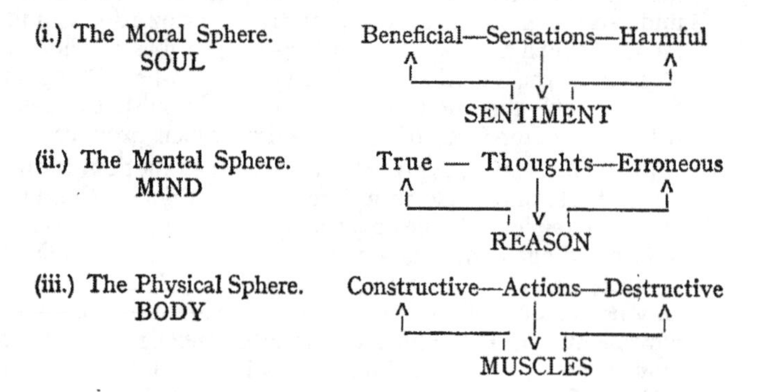


The problem throughout life is how to control these three spheres.

### 6. The Threefold Order of Man's Activities

I have now extracted from the organization of man three abstract quantities, or elements—namely, stability, activity and co-operation; and from his nature, three spheres of force—the mental, the moral, and the physical. In these three spheres the elements are ceaselessly at work, spinning as it were the life of the individual. I will now enquire into this phenomenon.

We frequently hear the assertion made that man has a right to live. In spite of the humanitarians, natural man, I hold, has no right to live, but, possessing power to protect his life, his might becomes the right to safeguard it. This power is manifested through movement, so once again we find a threefold order, namely:

(i.) Desire to protect life.
(ii.) Power to work or to fight.
(iii.) And ability to move.

The first is man's stable base, the second his active, and the third his co-operative, element.

Possessing power to move, he is enabled to work, and, through work, to protect his life by supplying himself with food, warmth, and shelter.

Whether we examine man as a highly cultured being or as a primitive savage, we find these elements in constant operation through co-operation, always present, and only varying in degree. In highly civilized communities work takes many forms, mental as well as physical, altruistic as well as egoistic, but it still remains work. Social rights are evolved from customs, and, to the common eye, a moral right to live is established, and yet is safeguarded by the power behind this right as manifested in the law and the police and soldiers behind the law. Thus, if we examine the structure of even the most highly civilized society, we shall find that moral power is based on physical power, just as it is in man. Further still, that moral power is established as a means of economizing physical power, so that human activity is not only expended in safeguarding the individual, but in securing the community, as well as increasing the general prosperity of peace.

From the individual man I will now turn to a group of men—a tribe, community, or nation. Here we find no radical change, only a difference in degree.

In a primitive society each man has to work for himself, and he carries a weapon to protect himself, consequently the rise of culture is slow, as the nation is literally a nation in arms. It is here that the establishment of a moral right comes to his assistance. Man has to work and to fight, but the less frequently the workers and fighters coincide the better it will be for the community as a whole, and the better it is for the community the better it is for each individual composing it.

The community, or State, as an abstract conception, stands between work and fighting, and manifests in the form of order. We thus obtain three national elements:

(i.) Protection, which is the stable element.
(ii.) Industry, which is the active element.
(iii.) And tranquillity, which is the co-operative element.

The first is the basis of military power, the second of economic power, and the third of ethical power. These are the three great political forces of a nation, of which military power is the base of all action; for by this power law and order are enforced, taxes are collected, communal expenses are paid, and the tax-payers, being freed from protecting themselves, can expend their energy on fostering prosperity, and the community as a whole is safeguarded against invasion.

The more prosperous a nation becomes the larger can be its armed forces; and the stauncher is the will of the people the more powerful do they grow. We thus see an intimate relationship between the nation and its fighting forces, which grows closer and closer as national power expands. The link between these two is government. Thus we get another expansion of the threefold order. During peace-time the armed forces are the stable element and the nation the active, and during war-time it is the reverse, for then the nation becomes the base of military action. Meanwhile, during both these periods, the government is the co-operating link which endows the one or the other with an increasing or decreasing mobility.

As primitive society is based on brute force, so also is civilized society, for armed force not only secures the nation against internal discord and external injury, but it enables its government, during peace-time, to enforce the will of the majority of the people on the minority, and also on foreign nations, by a threat of the application of physical force; consequently we find that an army is possessed of a threefold purpose:

(i.) It maintains domestic tranquillity by force.
(ii.) It maintains national security by force.
(iii.) And the link between these two is moral persuasion through the threat (and ability) to apply force.

I have now established, or attempted to establish, three leading ideas. The first is that man himself is organized on a threefold order, the second is that he is the product of a threefold force, and the third is that his activities may be summarized in three great divisions. We thus obtain a human instrument charged with power which is expended profitably or unprofitably, according to the object in view and the degree of knowledge possessed in its economy.

With a nation it is the same; for the society which man creates is but a development of his threefold organization, nature, and activity in a higher and more complex form.

In this society armed force finds its place, and, drawing its power from the nation itself, it consequently stands in close relationship to all the national activities, and through them back to the threefold organization of man.

I will now examine this relationship in order that it may be seen where armed force enters into the national scheme.

For a moment I will return to man. He has a soul, mind, and body, interwoven and interfused. In a crowd of people the mind, as a controlling organ, ceases to operate, and the soul of each individual merges into what may be called the spirit of the crowd; instinct, in fact, replaces reason. To obviate this chaos, a nation either submits to the will of one man or to a body of men directed by one man; thus a political control is established which regulates the relationship between the body and soul of the nation.

Thus, if the idea of a crowd of men is replaced by that of co-ordinated national power, this power may be divided into a threefold order. From the national body is derived the economics of the nation, from the national soul its ethics, and from the national mind its politics.

Diagramatically this may be shown as follows:

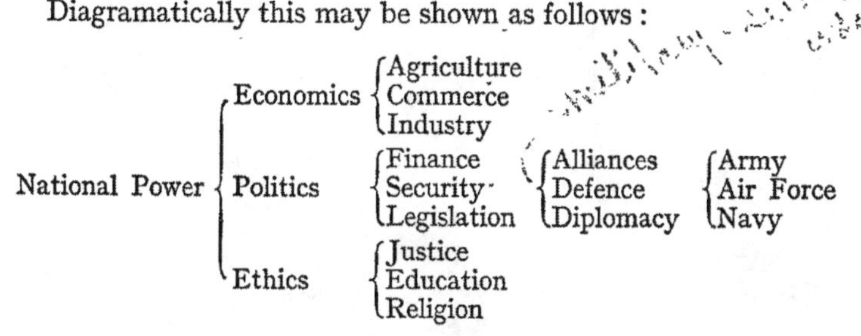


Security is the pivot around which the whole system revolves. It guarantees finance, which links government to national economics; it also secures legislation, which links government to national ethics. It is not purely military, for alliances and diplomacy are bracketed with defence. The defence section is the one which mainly concerns us, and to-day it is threefold, since air-power has been added to land- and sea-power. Each of these three major arms is composed of men, and in war each is ultimately controlled (or should be) by one man, all the remaining men being the vehicle which expresses the will of their respective commanders, who draw their inspirations from their government, which co-ordinates and controls the power of the nation. A nation built up of human cells, each slave to its instincts, and yet controllable through its faith, which is child of its knowledge and beliefs.

# CHAPTER IV

## THE OBJECT OF WAR

The legitimate object of war is a more perfect peace.
—GENERAL SHERMAN.

The legitimate object of peace is a more perfect man.
—ANONYMOUS.

### 1. THE FOUNDATION OF WAR

THE world is not governed by reason, but by the law of causation, or of uniformity; that is, similar causes produce similar effects. Without this law, which in itself is an assumption, as all laws formulated by the human mind must be, the scientific method would be impossible, in fact, it would possess no base wherefrom to operate.

If the student will now turn to the opening section of the last chapter he will see that the mind working within the trinity of space, time, and force realizes its surroundings in the forms of knowledge and belief, and that the intensity of either of these realizations constitutes faith, or the intellectual egoity of the subject. If this faith is firmly based on a close relationship between law and objective facts, it assumes a scientific character, but if on a relationship between assumption and subjective longings, then an unscientific one. The alchemical attitude is, as I have shown, a half-measure between these two, for it is a mixture of subjective desires and uncorrelated objective facts.

Turning now to warfare, I will substitute war for space. War is the area in which the soldier must work, and the history of war may be compared to time and military power to force. History is the record of time, or, rather, of the events which take place in time; it has its past and present, and, speculatively, its future. Military power, like force, is a compound of mass (body) and energy (activity), which expresses itself in the form of movement throughout its three spheres—the mental, moral, and physical.

In this new trinity I will place the mind, and we at once see that its operations are similar to those obtained in the original trinity.

By observing the facts of war, not only as they go to build

up military power, but as they have gone to build up military history, and continue to build it up at the present moment, we obtain knowledge of cause and effect. At first our premises may be hypothetical—that is, we believe that some theory is correct, or subject to proof; secondly, we actually prove it, and only accept the result when we are as certain as we can be that our reason for acceptance is no longer subject to exception. Such reasons constitute true military faith.

This, then, is the difference I am attempting to establish between the system I am now expounding and most of the systems which have preceded it: My military faith is based on an examination of facts correlated by the scientific method; the faith of the military schoolmen is based on unexamined, or badly examined, facts and assumptions. The struggle is between the adherents of two faiths, consequently it is likely to be a long one.

## 2. The Biological Causes of War

When the man of science has established a relationship between cause and effect, and has thus given expression to a reason, he is in possession of a fact worth knowing. The soldier, if he aims at working scientifically, must follow suit, and the first fact he must establish is the cause of war; for the cause of a war will produce its effect, not only during the war, but in the peace treaty which will follow it. Unless we understand the causes of a war, it is unlikely that we shall, from the outset, be able to formulate the object of the war, the attaining of which will lead to the effect required.

In human affairs it is mind which replaces law, and, though mind and law should be correlated, we cannot doubt that mind possesses freedom of choice—that is, it can disobey laws as well as obey them; and between these two, obedience and disobedience, lies the entire sphere of life as we know it. By obeying we utilize, and are rewarded; by disobeying we waste, and are punished—punishment is the measure of our error. Let us, therefore, obey, and obey knowingly and not blindly, for blind obedience is to reduce ourselves to the position of a stone unconsciously drawn towards the centre of the earth by gravity.

Fights are the concern of individuals and small groups of people; wars are the concern of nations; yet wars are built up of fights; consequently I will examine the causes of war, first from the standpoint of the individual, and secondly from that of the nation.

The strongest instinct in man is that of self-preservation, and I am of opinion, as I stated in my last chapter, that, because of this instinct, man possesses a natural and indisputable right to protect his life, not on moral, but on physical grounds, because he possesses the might to do so. This instinct is the keystone in the struggle for existence, which may, I think, be accepted as one of the main causes of evolution. To mitigate this struggle mankind establishes moral conventions and rights, but in wars for existence these conventions are set aside, and the contending nations become primitive savages, using the whole of their might —physical, moral, and mental—to preserve their national independence.

From the outset a point I want the reader to realize is, that in this struggle there is no essential difference between peace and war. The differences are purely relative. The essential is that might, or human energy, "demands action"; all action is struggle, and "every action is a conflict," and, as one writer says: "To put an end to conflict is impossible. Life is a conflict. As long as it lasts conflict will endure."[1]

To return to man. Another writer tells us that "Children do not fight because they are teased, they tease in order to fight,"[2] and a little observation will assure us that this is generally true. The same author writes: "Fighting play, therefore, prepares the young animal, not to attack feebler species which are to serve as his food, nor to resist stronger which covet him as prey, but, above all, to measure himself against other individuals of his own species"; because "It is to struggle for a female, rather than for food, that the young are being unconsciously rehearsed. . . ."[3]

If this statement be accepted as correct, then there is not only what I will call a military cause for fighting—that is, self-protection —and an economic cause—the search after food—but also a biological cause—the survival and improvement of the race. Turning to national life, the normally healthy nation does not only fight another to exterminate and plunder, or to prevent itself being exterminated and plundered, but to establish or maintain its ideal state of peacefulness. The animal man fights for a mate, the social man for peacefulness. Woman rears the family, peacefulness rears the State. The biological cause thus passes into the ethical cause—the maintenance of peace—and the same energy which is expended in the establishment of peace is utilized to preserve and to secure it. I think, therefore, that William James is right when he says:

> Every up-to-date dictionary should say that "peace" and "war" mean the same thing, now *in posse*, now *in actu*. It may even

[1] *Courage*, Charles Wagner, p. 193. [2] *The Fighting Instinct*, P. Bovet, p. 53. [3] *Ibid.*, pp. 45, 46.

reasonably be said that the intensely sharp competitive *preparation* for war by the nation *is the real war*, permanent, unceasing; and that battles are only a sort of public verification of mastery gained during the "peace" intervals.[1]

We thus obtain three fundamental biological causes of war: security of life based on the instinct of pugnacity; maintenance of life based on the instinct of hunger; and continuity of race based on the instinct of sex. The first is the mainspring of the military cause of war; the second of the economic cause; and the third of the ethical cause.

### 3. The National Causes of War

If the reader will now turn back to the final page of the last chapter he will see that these causes of war are closely related to the threefold order of national power, the only difference—and this is purely one of degree—being that in an organized nation military power is replaced by political power. As I say, the difference is only one of degree, for political power, just as much as military, is based on brute force, the ballot taking the place of the bullet.

From the three spheres of national power emanate three great groups of causes of war. We have at first those of race, of education and religion, which give us ethical causes; secondly, those of commerce, industry, and supply, which lead to economic causes; and thirdly, those of geography, communications, and fighting strength, out of which evolve military causes.

Racial causes are ever present, and yet are difficult to fix. Accepting nations as great groups of individuals, a more pronounced hostility exists between them than between the individual members of each group. In Europe, for centuries we watch an undying enmity between Teuton and Latin and between the Nordic and Mediterranean races, due, no doubt, to the fact that their psychological outlook is different. These racial differences are accentuated by religion and education, for, whatever the origin of a religion may be—and most are Oriental, and consequently foreign to European culture—in place of assimilating race psychology they are assimilated by it, until out of one root can sprout three such different trunks as the Catholic, Greek, and Protestant Churches.[2]

Economic causes are also fundamental. Each nation, like each individual, desires prosperity, and if a nation be strong it

[1] "The Moral Equivalent of War," in *Memories and Studies*, W. James, p. 273.
[2] The cradle of a nation is frequently an internal religious war.

will attempt to gain it. In former days plundering was a cause of war, now it is commerce, and the difference is again only one of degree. The acquisition of undeveloped lands in order to obtain raw material, the control of markets where manufactured goods can be profitably sold, and the command of communications, especially those of the sea, to assure the safe passage of raw and manufactured materials, are all potent economic causes of war.

Possessed of a high ethical and economic power, a virile nation very naturally determines to secure itself from either internal or external interference. This search after security is the most potent of the military causes of war. Internally, during peace-time the nation is an entrenched camp. The will of the majority, enforced by the national Government, maintains a state of peacefulness by force, for this will is backed by military power. Externally—that is, against neighbouring or competing nations—this will can only exert its power indirectly by threat of force, and when two nations threaten each other, however amicably, the desire for security leads to the search after strong or unattackable frontiers. I will take a simple example.

A man, before retiring to rest, bolts the windows and locks the doors of his house, and, if he lives in a lawless country, he may place a revolver by his bedside. The outside walls of his dwelling are his frontiers, the bolts and locks are the fortresses blocking the natural avenues of approach, and his revolver is his field army. From the individual I will turn to the nation. The stronger its walls and frontiers are, the securer it will be. If they are weak, fortresses and field armies must be increased. The wise man builds a strong house, so also does the wise nation, and if the nation be powerful, and yet possesses weak frontiers, it will seek to strengthen them as surely as a rich man will refuse to live in a barn if he can obtain a brick mansion. This, then, is the point we must grasp: every healthy nation which possesses the power to establish strong frontiers will attempt to do so, either by occupying natural features which will strengthen them, or by creating weak neighbours who dare not cross them. An examination of history will show that this is so, and that the search after strong frontiers in order to secure peacefulness is a fundamental cause of war.

These three great groups of causes produce their effect through political action which, by concocting a pretext, detonates the war. In wars between great democratic nations it is the nations themselves, and not their Governments, which are responsible for war. The politician may hasten or retard the outbreak of a war, but unless the causes are potentially in the soul of the nation a great war is impossible.

This is the point which is nearly always missed or glossed over by pacifists and humanitarians. Because in domestic affairs the ballot has replaced the bullet as a means of expressing force, they assert that a similar moral equivalent for war can equally well be established between nations. In this assertion there lurks a deadly fallacy.

In all democratic countries the might of the majority makes right. No court of justice can reverse the decisions of the ballot-box, for such a reversion is only possible through the will of the majority, or a revolution in which the minority succeeds in imposing its will on the majority. Whilst in a nation a moral equivalent for war has been discovered, none has so far been found between nations. Arbitration cannot settle international political questions of importance. Because no court of justice can settle political questions within nations, so equally can no court or commission settle international political questions between nations. As Colonel Vestal says:

> You will find that in every nation in existence to-day the right to declare war is lodged, for all practical purposes, in a body which has power to raise and support armies and navies and to raise revenue to carry on war. . . . You can never take from the Congress of the United States its power over the sword and give it to an international body, unless you give the international body the power to tax us to pay for making war. Manifestly we will never do that. . . . If it were possible to establish an international legislature which had power to make war and unlimited power of taxation, the ballot would, of course, become the moral equivalent of war for settling political questions in the world state. The most enthusiastic internationalist, however, has never proposed a real legislative union of the world.[1]

To-day, from their major point of view, Leagues of Nations are leagues of nonsense, as they cannot control the causes of war. The only factor which throughout the course of history has done so with any success is what is called the balance of power, which aims at meeting pressure by resistance. In the past, this balance has only been completely upset when the aggressor has simultaneously possessed command of the land and command of the sea. How far command of the air will complicate this balance I cannot discuss here, but the past tells us this—that as long as one power is supreme on the sea and another on the land the conquest of the world—or known world at the time in question—is not a feasible operation.

[1] Lecture given in February 1923 on "The Maintenance of Peace." See also Colonel Vestal's book, *The Maintenance of Peace*. Curious as it may seem, such a union was the ultimate aim of German world-power in 1914.

## 4. The Object of War

From the causes of war I will now turn to its object, aim, or purpose. First it should be realized that its object is closely related to its causes. In its most condensed form the cause of war is discontent with the existing conditions of peace, but, as the nature of peacefulness is complex, so out of this one cause, as I have shown, evolve three great groups of causes, and, when once war is declared, each of these groups is confronted by a correlated group of objects, the gaining of which will remove the discontent which has led up to the war.

The object of a nation as a self-governing unit is prosperous racial survival, and to all individual and family requirements must be added the need of co-operation between individuals and families as well as self-sacrifice for the common or co-operative good. For a nation to survive it requires :

(i.) Self-sacrifice leading to ethical superiority (culture).
(ii.) Control leading to political stability (order).
(iii.) And co-operation leading to commercial prosperity (comfort).

The three, conjoint, constitute the means of maintaining the object of a nation which, when given expression through its Government, constitutes its policy, the maintenance of which is the object of political control.

In order to maintain, protect, and enforce policy, all civilized countries raise armed forces, the object of which is to maintain domestic peace and to secure the nation against foreign invasion and diplomatic threat.

As the policy of a virile nation is to enforce its will on its antagonist, the sooner it can do so the less commercial capital will it expend, and the less disorganization of existing markets, whether in its own hands or in those of its enemies or allies, will result. In wars originating in economic causes the object is not to kill, wound, or plunder the enemy, but simply to persuade him, by both moral and physical pressure, that acceptance of this policy will in the end prove more profitable than its refusal; for to kill, wound, and plunder is to destroy or debilitate a future buyer—it is, in fact, a direct attack on the competitive impulse which is the foundation of prosperity.

From wars arising from military causes, frontier security, etc., it is much the same. The object is to remove the military threat with as little injury to the hostile nation as is compatible with its attainment. In wars arising from ethical causes, such as the loss of independence, of ideal, or of religious freedom, unfortunately it is otherwise, for the objective aimed at is intangible; it is not a frontier or a market, but an idea ; hence it happens that the most ferocious of all wars are civil wars and wars originating from religious causes.

To return to the object. The nation replaces the man, its ethical outlook—the soul : its economic wealth—the body ; and its political system—the mind. We thus obtain a close coincidence between the nation and its ultimate units—the men and women who go to build it up. The object of man is to live, and to live contentedly and prosperously ; similarly, the object of a nation is to exist, and also to exist contentedly and prosperously. The brain of man is his controlling organ ; so also is the Government the national organ of control. Diagramatically we obtain the following :

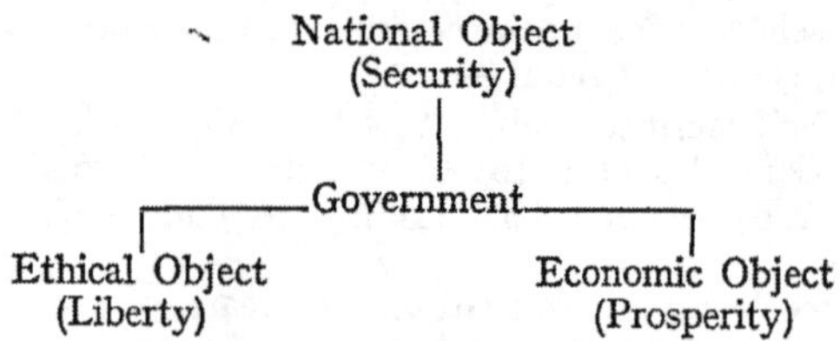


I will now consider these three objects from the point of view of war.

## 5. The Nature of the National Object

Both in peace and war, the backbone of a nation is its racial character. This backbone supports its civic body and forms the base of operations for its military limbs. In war, as in peace, the character of the nations competing form the foundations of their policy, diplomacy, and effort. Character is the sun which lightens the whole horizon of endeavour ; glowing with racial instincts, its rays are received, refracted, or obscured by local customs and traditions, which lie deeper than intellect or reason. In normal circumstances its full powers remain eclipsed, and they are, consequently, difficult to appreciate, but as it is so often the event which reveals the man, so also, in great national crises such as war, the character of a people assumes its full and inherent form, and manifests as the light and leader of the nation.

This is undoubtedly so, consequently it is during great wars—struggles for existence—that character attains its most tangible form, and reveals itself in the will to win or to accept defeat. If the war be unimportant, its loss may not materially affect the nation ; nevertheless, it will be a blow registered against its prestige, its moral capital, on which so much of its material prosperity is based. Its credit will be lowered in the eyes of others, and a series of such blows may exhaust the national *moral* to such an extent that the will of the nation is laid bare to a knock-out blow.

If the war be important, victory becomes vital, and the nation, subconsciously realizing this, sets about to divest itself of the formalities of everyday life. Traditions, customs, and party aims are, one by one, discarded and replaced by common sense actions, and, as this process grows, the great static and foundational racial spirit reveals itself, and a nation, according to its character, stands or falls.

National solidarity is a psychological and not a physical phenomenon; further, wars between democratic nations are not originated by pushful or piqued individuals, but by the nations themselves. It is, therefore, the nation which is the true aggressor, its Government being but its trumpet. It is the national will to win which must be broken, consequently it is this will which forms the basic military objective in war, the object being its conquest.

Once this will is broken the war is won; but, in the breaking of it, it must be remembered that the enemy's Government should not be bereft of its domestic powers, or else the enemy will be bereft of his national brain. The attainment of the national object aims at an agreement and not at a mental disruption of the hostile nation. To reduce a nation to a state of idiocy or of anarchy only means that it will be deprived of the power of fulfilling its contract—the terms laid down in the peace treaty. And if these terms are not fulfilled, then, from the point of view of policy, the war will, to a great extent, have been fought in vain; for policy should aim at attaining a more perfect peace than the one unhinged by the outbreak of hostilities. Conversely, the contract must be reasonable; for to compel a beaten foe to agree to terms which cannot be fulfilled is to sow the seeds of a war which one day will be declared in order to cancel the contract. Thus the national object is a better peace, and the means of attaining it is the conquest of the will of the hostile nation.

### 6. The Nature of the Ethical Object

The attainment of a better peace demands a higher ethical outlook. This brings me to the ethical object of war, which is the enhancement of the national character—to increase its prestige, not only in the opinion of the enemy, but in that of all other nations. A man who fights cleanly is always applauded, even if he loses; consequently, in certain circumstances it is even more important to win the ethical object than the military object. To be proclaimed an international cad in the world's opinion is equivalent to being regarded as such in the public eye.

Chivalry, in the broadest meaning of the word, is the cultivation of respect in an enemy for or by his opponent. Outstanding acts of courage, of courtesy, and of humanity give birth to a feeling of superiority or inferiority, according as one side excels or falls short of the other. This feeling of superiority, of *noblesse oblige,* is purely ethical, yet it forms the foundation of the physical superiority which war demands. The side which first attains a superiority in chivalry is the side which attains a moral victory over its enemy—a victory which frequently not only precedes physical success, but which wins the ethical object of the war, which is the true foundation of the peace which follows it.

War in many respects is comparable to a game. It has its rules, which are elastic enough to be of general application; but there is this difference, that whilst in a game the referee is represented definitely by a third party, in war he is only represented by the conscience of the combatants themselves as influenced by the ethical opinion of neutral States. In wars other than world wars this opinion has a profound influence on the behaviour of the combatant nations, but in world wars it ceases to hold sway, since no nation of importance remains neutral. The referee removed, the result is that the war rapidly develops into a cad's struggle, the low ethical tone of which becomes clearly apparent in the peace treaty.

Though in wars of all types there is no belt which may not be hit below, nevertheless a wise fighter will think twice before hitting below a certain moral line, because the material advantage accruing may be cancelled out by the ethical loss resulting.

These high ideals must not, however, blind us to common sense. Men who take on the nature of vermin must be exterminated, and in their extermination the entire moral progress of mankind is moved one step nearer to its final and unknown goal. To refuse to use brutal means against a base foe is to set a premium on crime, and in war there are crimes as well as honours. To tolerate crime is neither to act chivalrously towards a criminal nor chivalrously towards oneself; it is the act of a fool—that is, of a man who values his self-preservation at the price of a custom which, ceasing to be marketable, has become counterfeit.

Ultimately it must be remembered that, on account of the intricate economic relationships existing between civilized nations, great wars are becoming more and more world wars, and as the victor in a great war will, in the peace which follows final victory, exert a higher influence on civilization than the vanquished, it is an advantage to the world as a whole that the cleanest fighter wins. Consequently, to fight cleanly is to be supported by what is righteous in the world's opinion.

### 7. The Nature of the Economic Object

In its ultimate form the economic object in war is the national object, namely, survival with profit, which presupposes an ethical outlook, since honesty endows prosperity with its firmest foundation. If this objective is to be attained in a full degree, then the peace which follows a war must at least be as prosperous as the peace which preceded it, for prosperity is the material dividend of victory.

I can, I think, explain this more clearly by returning to my example of a duel between two men. Economically, it is not sufficient for the victor to kill his opponent, for he must secure himself against being so badly mauled that at the conclusion of the struggle he is left permanently crippled. Further, should his opponent be his buyer, and should the quarrel have arisen over a question of barter, economically the objective will not be gained by destroying his adversary, for this very act will defeat the end in view. Rather should it be sought for through disarming him, which will enable such terms of peace to be dictated as will compel him to sell and buy at values which are economical to the victor.

If a man be fatigued or in poor health his muscular endurance will be low, he will be lacking in staying power; if the reverse, his staying power will be high, for it will consist of that surplus of muscular energy which is not actually required for the maintenance of his daily existence. The amount of this staying power can never be excessive, and the skilful fighter, knowing this, is most careful in its expenditure; in fact, he realizes it economically; that is, he attempts to spend less energy in proportion to his efforts than his adversary, and yet by doing so gain equal, if not superior, results.

To-day industrial endurance forms the staying power of war, and, as it can never be excessive, a wise Government should see that during war this wealth is squandered neither by civilian nor soldier, and that war expenditure is remunerative in the fullest meaning of the word, namely, that it could not have been more profitably spent.

As in the individual the staying power for war is measured

in terms of surplus muscular endurance, so in the nation is it measured in terms of financial endurance, which represents the surplus productivity of the nation's work. Accumulated wealth or money has, therefore, been rightly termed "the sinews of war," and if this be realized it will at once be seen that the economic object in war does not only consist in destroying the enemy's strength, but in destroying it with profit. If this is done, then peace will find the victorious nation in a superior position to that in which it was on the declaration of hostilities. It can then not only gain an advantage over the vanquished, but can compete with all other nations.

On first thought it may be considered that this is not a question which concerns the soldier, but solely the financier and politician; but, on second, I think it will become readily apparent that, unless the soldier understands the true meaning of the economic object, he has no right to complain if politicians and financiers, and these two always run in harness, attempt to direct a campaign so that its cost does not permanently cripple the nation.

### 8. The Political Object

These three objects—to exist, to exist honourably, and to exist profitably—are, or should form, the directing forces of political power. A nation, like any crowd of individuals, is inarticulate without a leader or a national assembly, because it is controlled by instincts and not by reason. Its government, whatever form it may take, is its thinking organ, drawing its sensations from the nation, and converting these into reflections, and from reflections into decisions, and, lastly, actions.

Unfortunately, to-day, governments generally work on lines just as alchemical as military organizations; and, though innumerable books on political science have been written, governments do not carry out their work on scientific lines. In place of mastering their environments, they, more often than not, are mastered by them, and especially so if these environments are those of war.

Lord Morley once said that politics were neither a science nor an art, but a dodge. This is very true of politics to-day; consequently, when in war, the military alchemist is controlled by men whose upbringing has been one of dodging difficulties in place of conquering them, the result is frequently disastrous.[1]

[1] Edward III, in 1372, to facilitate parliamentary procedure, forbade the election of lawyers; in 1404 Henry IV did the same, and the result was the "Unlearned Parliament," which justified the King's action, as it got through a great deal of work.

If, as I have attempted to show, it is necessary for soldiers to understand the nature of the causes of a war, since these are closely related to its objects, how much more so is it necessary for politicians to understand them, since they represent the national will which so largely creates these causes. This understanding or misunderstanding, as the case may be, is expressed consciously in the policy of the government. Policy is, in fact, the relationship between will and surroundings expressed in words. On one side of the politician stand the esoteric instincts and desires of the nation, and on the other the exoteric facts of life—these it is his duty to correlate.

Domestic policy, *per se*, is the national purpose derived from the correlation of all the qualities and quantities which go to build up the national, ethical, and economic objects, but it never can be considered *per se*, since each nation is part of the world, and to-day, on account of the interfusion of ideals and of wealth, not only a national but an international part. Whatever influences a great democratic nation influences the whole democratic world, mentally, morally, and physically. We no longer live in the period of isolated national shocks, but of ceaseless international repercussions. Thus we find that domestic policy must, in its turn, be correlated with the policies of all other nations—hostile, neutral, and friendly—and that out of this grand correlationship springs foreign policy.

In the main, the object of policy is first to maintain and enhance the general prosperity of the nation, and secondly to secure it against internal and external interference. The problem of war is, consequently, always present, and the political object in peace or war is a more perfect peace. If this object is not attained, then, though the war may not have been fought in vain, it will not have fulfilled its highest purpose, which is to create a better state of conditions, and not merely to destroy an existing discontent.

Power to wage war should, therefore, be looked upon as a creative force, and not merely as an insurance against calamity. To-day this outlook on war scarcely exists, and, in my opinion, it will never exist until a science of war has been established, by which the conception of war may be correlated with our conceptions of all other human activities.

All honour is due to Clausewitz for having made clear the relationship of policy and war. "We maintain," he writes, ". . . that war is nothing but a continuation of political intercourse with a mixture of other means."[1] And again: "We see, therefore, that war is not merely a political act,

[1] *On War*, Von Clausewitz, vol. iii, p. 121.

but also a real political instrument, a combination of political commerce, a carrying out of the same by other means . . . for the political view is the object, war is the means, and the means must always include the object in our conception. . . . State policy is the womb in which war is developed, in which its outlines lie hidden in a rudimentary state, like the qualities of living creatures in their germs."[1]

Yet what little attention does the politician give to war, a force which everywhere surrounds him, and which any one of his actions may render sensitive to explosion. It is amazing to contemplate this ignorance, which, as democracy advances in power, becomes denser and denser, and so dense that the world must inevitably be engaged in unrighteous war. I will, therefore, lay down certain economic rules or maxims as guides[2] to those who wield political power as if it were a harmless combustible.

Granted that the object with which nations go to war is to attain better, or to ensure against worse, conditions, then the loss of life and capital is compensated for, not by military success, but by the attainment of this object through military effort. Though it may often happen that military success can only procure the desired conditions of policy or stave off the undesired ones, it must not be forgotten that it is only as a *means*, and not as an end, that it is of value, for wars waged otherwise must normally prove uneconomical. This holds good whether the war be offensive or defensive in character, for even if defensive, though the object is not to enforce a policy, it is nevertheless to safeguard a policy the aim of which is to maintain national liberty and prosperity.

From this we may deduce the following, namely, that:

"*A military victory is not in itself equivalent to success in war.*"

What is equivalent to success is a more prosperous peace following the war, and though this condition may seldom be attainable, yet it constitutes an ideal worth striving after.

War not being an end, but a means, the financial situation at its conclusion must be considered coincidentally with the results of military victory in so far as they effect the future well-being of the country. Every man killed means a loss of capital. Every shilling expended is a mortgage of a shilling's worth of production after the war. Wages and prices are thus adversely affected to a definite and calculable extent by each day's operations.

Again, loss of capital resources on the part of the enemy cannot

[1] *Ibid.*, vol. i, pp. 23, 121.

[2] The following principles are based on a paper on *War Economics*, written by Brigadier-General Ramsay Fairfax, C.M.G., D.S.O., late Royal Navy and Royal Tank Corps.

figure on the credit side of our account; hence the defence of lavish expenditure as leading to the war bankruptcy of the enemy is unsound, seeing that the enemy is a potential buyer; and, consequently, to destroy him so utterly that he ceases to possess the power to buy, is to deny ourselves a profitable market, and so strike a blow at our national preservation. Therefore:

"*A war, to be economical, must enforce acceptance of the policy under dispute with the least possible harm to commercial prosperity.*"

Accepting these conclusions, the value of military success decreases in proportion to the total expenditure, and from this it follows that there exists a theoretical limit of expenditure, on exceeding which military success ceases to be on the balance profitable; consequently all operations not contributing directly to a decision shorten the time available in which it may profitably be sought. It follows then that:

"*A military decision, to be economical, must attain more profitable result than the depreciation of capital due to its attainment.*"

From this it follows that unless each operation contributes to the final victory, in proportion to its cost, it shortens the time available and diminishes the value of eventual victory, or hastens defeat.

The whole of this process of arriving at an economical war policy throughout the history of war has been conspicuous by its absence. In itself it is a science, yet it has never been treated as such; hence the general chaos of war.[1]

The whole of this question of the formulation of war policy is too immense for me to deal with in this book, but I hope that I have dealt with it sufficiently to accentuate its importance. War policy is the continuation of peace policy. During peace-time the power of the government is founded on the national will, and the instrument of the government is national force, of which part is called military power. In war it is the same, and the only remarkable difference is that, whilst during peace-time danger is absent, military power compels the minority to accept the will of the majority. A national danger, threatening majority and minority alike, cancels their differences and enables a government to turn military power outwards, and so compel the enemy to accept this same will in its full national form. War, and not peace, is the true condition which gives expression to nationalism.

[1] In peace-time the object of a government is secure and contented prosperity. This object is based on certain factors; these factors must not be destroyed in war.

# CHAPTER V

## THE INSTRUMENT OF WAR

Force rules the world still,
Has ruled it, shall rule it;
Meekness is weakness,
Strength is triumphant
Over the whole earth.

—LONGFELLOW.

### I. THE POLITICAL INSTRUMENT

CLAUSEWITZ considered that war was not merely a political act but the real political instrument.[1] I have no quarrel with this assertion; nevertheless, I prefer to look upon war as the condition resulting from a more strenuous and concentrated application of force to the normal political instruments used in the maintenance of peace. In brief, during peace-time tranquillity is established by law and order, which is maintained not only by force, but by a regard for individual liberty and a just distribution of wealth. Force is always present, but in a well-balanced country it is kept out of sight. In war force steps to the front, and hitherto has been the main political instrument to compel an enemy to accept the will of the nation.

From the highest aspect of this subject, the nation itself is the political instrument, but as, outside its government, it possesses no co-ordinated mental power, the government is the craftsman who makes use of it, and, as the power of the nation is threefold, the political instrument is threefold in form. The government can bring economic, moral (ethical), and military force to bear against its enemy. It can directly, through political action, bring economic and moral pressure to bear by means of financial and commercial restrictions and by propaganda.[2] It

[1] Sir Walter Raleigh considered war the failure of political action, rather than its instrument.

[2] The value of propaganda was much exaggerated during and after the Great War of 1914–18. Lies nearly always recoil on the head of the liar; and most of British propaganda consisted in the kettle calling the pot black. The force of true propaganda lies in its truth, as truth is so often allied to fearlessness. A nation, or man, who is not afraid of hearing the truth is of high *moral*.

can also indirectly attack the will of its adversary by means of its fighting forces.

I do not intend here to examine the purely political activities of war, not because they are unimportant, for they are of ever-increasing importance, but because they form subjects concerning which I am not well acquainted. I will therefore in this chapter concentrate my attention on the organization of the military instrument.

## 2. The Military Instrument

In chapter iii. of this book I accepted as my architypal organization the body of man, and, by examining this organization, I came to the conclusion that it revealed a threefold order of structure, maintenance, and control, and that each of these factors was built of three elements—stability, activity, and co-operation. These facts—and I think that they are true facts—I will accept as my present base of action.

The military instrument is man, or a number of men. To-day, in most highly organized nations, it consists of an army, a navy, and an air force; its military power is consequently a threefold one, for its force can be expended on land, at sea, and in the air. Until recently war space was two-dimensional, to-day it is three. We have arrived, therefore, at a close agreement between war and the conception of space itself.

Man moves in three dimensions, so to-day does the military instrument, the three Services of which in themselves do not necessarily give structure to the whole, since they constitute the "material" out of which structure is designed. This design depends on the relationship between these three Services and the conditions which are likely to confront the nation in war. In the past our naval strength has been the base of our military action, and as long as our military forces maintain their present organization this must remain so. How far air-power will influence military and naval organization it is difficult to say; and it is not here that I intend to seek a solution to this problem, since my immediate object is to accentuate the importance of structure in the military instrument, and not to examine the activities springing from it. The main point is that a highly organized nation has two or three fighting Services; consequently, if the structure of the military instrument is to possess a high stability, then the proportional strengths of these Services and the nature of their separate organizations must form an articulated, co-operative whole. That is, they must fit together economically, and, if possible, as economically as the bones of the human skeleton. As Jackson says of an army: "The whole conspires in one purpose; for, though an army consists of many parts, it is only one instrument, constructed for the accomplishment of one design."[1] Similarly, on a larger scale, must the whole of the fighting forces of a nation be set together to accomplish one design.

Without such a disciplined instrument, control is next to impossible. If the bones of the human body were not so shaped that they formed an articulated skeleton the brain could not control the body, and, without control, structure proves not only useless but detrimental. For, though the wielding of the instrument demands skill on the part of the wielder, "it is necessary that the means, placed in his hands, be rendered capable of a uniform and systematic action, calculated to second his views in the direction of his force. For, it being from the perfection of the instrument in its primary movement that decisive effect results in application, an army, correctly organized and animated internally, has often been found to conquer without the aid of uncommon ability in the general; an able general has often been seen to fail in his designs from the mere defects of his instrument—that is, the want of harmony in its mechanical movement, resulting from an injudicious composition of the subordinate parts. Hence the primary organization of the materials of an army, supported by the discipline of tactic, is an object of great and essential importance in controlling events in war."[2]

This is not only an undoubted historical fact, which has proved itself time and again, but a very important fact, for, as it is not possible to assure command being carried out with genius, it is, nevertheless, possible to create a well-organized instrument. In the case of man, his organization has grown as a whole; it has not developed in parts and then been set together. Though with the military instrument the problem is not so simple, there is no reason why one man, or a committee of men, working scientifically, should not so design its parts that they will fit together in place of being stuck together. Of this Jackson says: "The direction of the action of the military instrument is under the management of the military officer; the organization of its parts and the adjustment of its powers is more peculiarly the work of the scientific philosopher. The fundamental arrangement requires a deep knowledge of the principles of elements, whether physically or morally considered."[3] This is most true, for, if this articulation is guaranteed, then, when it comes to

[1] *A Systematic View of the Formation, Discipline, and Economy of Armies,* Robert Jackson, p. 27. [2] *Ibid.*, pp. 138, 139. [3] *Ibid.*, p. 138.

war, it will be found possible to unify the control of the three fighting Services under one group of men, and eventually under the direction of one man, and so establish a complete command over the instrument.

Thus far, I think that the comparison of the military instrument to the human body is logical; equally so are the processes of maintenance, though at first thought this might not appear to be so. In the human body the organs of maintenance are internal; nevertheless, they are dependent on external supply. In the case of a ship they are internal, because mechanical power renders possible their carriage in whole or in part. But in the case of an army, depending on muscular movement, the organs of maintenance are so elementary that they have to be supplimented by an organization apart from the fighting body. Though external supply must always remain—since even ships cannot indefinitely be maintained at sea, and less so such mechanical arms as aeroplanes and tanks—the more the organs of maintenance are brought within the fighting body the more direct will be the action of this body, since the less will the protection of the administrative services have to be considered.

For a moment I will turn to the external aspect of this question. For example, if the instrument were to consist of, I will suppose, three men, each requiring different articles of supply, such as different rations, uniforms, tents, weapons, etc., the maintenance of such a force would be more complex than if all three required the same. So also if the military instrument consists of an army, navy, and air force, the more their maintenance can be unified the more easily can the whole be controlled.

It is not my intention to press this question, since my object is not to reform, or reconstruct, the military instrument, but, instead, to devise a piece of mental machinery which will enable any intelligent man to analyse existing military organizations and discover their defects.

### 3. The Structure of the Instrument

I will now examine in more detail the organization of the military instrument. Its structure is pre-eminently tactical, consequently its parts must be so set together as to enable its commander to develop its maximum fighting-power. In the case of two men fighting, the will of each is expressed by means of his fists. Each, if he is a trained fighter, protects himself with one arm and hits out with the other. The protection afforded by his left arm is the offensive base of action of his right. If his protection is defective, he may be thrown entirely on the defensive; that is to say, he may have to supplement his protective endeavours by means of his right arm. If he is strong and skilful, he may at times be able to supplement his right by his left. Whether he is driven back or whether he advances, the relationship between protection and offensive action is mobility—movement backwards or forwards, away from or towards the object directing his will.

Here we watch in operation the three elements of stability, activity, and co-operation in the forms of resistance, pressure, and movement. Foch, when discussing *Economy of Force*, describes how a general should "set up his forces in a *system* such that these forces may finally act in conjunction." His system is "a combination of the two qualities present in all troops," namely, "resisting power" and "striking power."[1] Lloyd is still more explicit; in brief, he says: "War is a state of action. An army is the instrument with which every species of military action is performed; like all other machines, it is composed of various parts; and its perfection will depend, first, on that of its several parts; and, second, on the manner in which they are arranged; so that the whole may have the following properties, viz. strength, agility, and universality; if these are properly combined, the machine is perfect. Care must be taken that not any of these properties is increased by diminishing another, but that the whole may be in proportion."[2] To Lloyd, strength is the collective vigour and weapon-power which enable an army to attack and defend, agility is quickness of manœuvre, and universality is to be sought in formation, which should permit of it moving against all kinds of troops and over all kinds of ground without changing its structure. He writes:

"The first problem in tactics should be this: how a given number of men ought to be ranged so that they may move and act with the greatest velocity; for on this chiefly depends the success of all military operations.

"An army superior in activity can always anticipate the motions of a less rapid enemy, and bring more men into action than they can in any given point, though inferior in number. This must generally prove decisive, and ensure success."[3]

I have inserted these quotations not only because they support my argument, but because they show how long it takes to establish a true fact. Lloyd, be it remembered, wrote his book about a hundred and fifty years ago.

Turning from the individual fighter to armies, navies, or air

[1] *The Principles of War*, p. 58.

[2] *History of the Late War in Germany*, part ii, p. 1. [3] *Ibid.*, p. 2.

fleets, we find action to be similar. The commander fights with two forces—a stable force which can resist pressure, and an active force which can exert pressure. These two combined, as Foch rightly says, constitute the foundations of tactical power, the commander making use of them just as an individual boxer does of his fists. In the case of the single man, should he wish rapidly to gain contact with his adversary, or escape him, he makes use of his legs. As it is not feasible to do the same with an army (I will now deal with armies only, as this will make the problem simpler), a third, or mobile, arm has to be introduced, which can operate from the other two, these two forming its base of action.

What do we see here? The expression of the three elements of force through three separate, though closely related, arms, or bodies of troops. Each of these arms, in its turn, in order to co-operate with the remaining two, should possess within itself stability, activity, and mobility, or, in tactical terms, protective, offensive, and mobile power. We thus obtain three main arms, each built round the three elements, and each expressing more fully than the remaining two one of these three elements. When structure is developed from these three, then tactics flourish as a high art; when it does not, then a period of decadence supervenes.

In illustration of the above, I will first examine the Grecian phalanx. Its combatants were divided into three main categories of soldiers: the light infantry, or *psiloi;* the heavy infantry, or *hoplites;* and the cavalry, or *cataphracti.* We here get a three-fold division of tactical power. The heavy infantry give stability to the whole organization; they form, so to speak, the bones of the phalanx. The light infantry operate from this stable base and demoralize the enemy; they can do so because they are more active than the heavy infantry. If the heavy infantry were to advance directly upon the enemy's heavy infantry, they could only engage on equal terms, or else, should the enemy retire, they will find it difficult to pursue, and still remain in an organized formation so necessary to withstand cavalry.

The light infantry can move quicker than the enemy's heavy infantry; consequently, if the hostile phalanx falls back, they can continue to annoy it at close quarters. If it advances, the light infantry retire behind their protective shell—the *hoplites.* They may not be so mobile as the cavalry, but they are more active, because their power of movement is assimilated more closely with their offensive and protective powers, whilst with cavalry it is separated from them, because the horse is not part of the man.

By annoying the enemy, the light infantry compel the hostile soldiers to protect themselves; that is, to stabilize their activity. The phalanx then catches up with the fixed enemy and breaks his organization into pieces. Eventually the cavalry follow and destroy the shattered fragments.

From these three elemental types of soldiers we see evolved three primary activities:

(i.) The light infantry demoralize, and, by instilling fear into the enemy, they fix him.
(ii.) The heavy infantry disorganize, and, by disjointing the hostile skeleton, they render it inarticulate.
(iii.) The cavalry destroy and, by stripping the flesh off the disjointed bones, they annihilate the enemy's resistance.

It will, I hope, be realized that this example is a very general one, for many battles have been won by light, or heavy, infantry, or cavalry, alone. But general though it be, the point I am out to accentuate is that the most economical military organization is one which expresses the closest relationship to the organization of the human body.

To continue the illustration. In the early Middle Ages infantry practically disappear, and, as cavalry are alone used, tactics become decadent. With the discovery of gunpowder, infantry reappear in full, and take the place of the old light infantry—the demoralizing agent. A new arm is introduced—the cannon—which carries out the protective duties of heavy infantry. All this takes, comparatively speaking, an immense time, for the only process of evolution is trial and error: Failure is the master, not forethought. Eventually we obtain the three arms as we know them to-day—artillery, infantry, and cavalry. The first forms the base of action of the second, and the second of the third. To-day, on account of the supremacy of fire-power, cavalry have largely lost their mobility, consequently tactics have once again entered a decadent stage, which was very noticeable during the Great War of 1914–18, for it was a war of tactical mediocrity.

I have entered into this somewhat detailed analysis with a definite purpose, namely, to show what constitutes fighting power, and not merely the type of soldiers who expend it. Artillery, infantry, and cavalry *are not necessarily essential arms*, because there is not such a thing as an essential arm. Arms are but means towards an end, and these means are constantly changing. What is essential is fighting force which expresses in full the three elementary powers. Wellington thought in terms of artillery, infantry, and cavalry, and not in those of pikemen, archers, and knights. Yet Edward III thought in these terms, and rightly so, since in his day these arms did express the elements of force. To-day we still think in the terms Wellington thought in, not because they express the highest forms of protective troops, combat troops, and pursuit troops, but because we fail to understand their spirit and can only grasp their names. In brief, we, or most of us, are obsessed by nomenclature, and are prejudiced through ignorance in the essential qualities of fighting force. Not until we overcome these prejudices shall we be able to think scientifically.

Finally, as regards structure, we arrive at the following conclusions: the structure of fighting force must be such that it will permit of the enemy forces being rapidly demoralized, disorganized, and destroyed, and, simultaneously, prevent the enemy carrying out these acts. Three types of troops are required, and these I have called protective troops, combat troops, and pursuit troops. These form the threefold structure of fighting force.

### 4. The Maintenance of the Instrument

Granted that the commander is the brain of the army he controls, then, to maintain its fighting force he must be prepared to make good deficiencies and injuries; in fact, he must supply his army and repair it. The first of these two requirements form the base of the second, for supply represents the stable element, and repair the active; and the link between these two is transportation, which expresses the mobile element.

On the one hand we have the structure of fighting force and on the other its maintenance. Obviously, these are closely related, since the second makes good the wastage of the first. The second is in fact the base of the first, and the more perfectly these two are correlated, the more fully can the control of the commander find expression.

If the structure of fighting force is such that supply is rendered difficult, however perfectly fighting force may be expressed, its endurance will be low, for it will lack staying-power. For example, in Japan there exists practically no automobile industry, and a very limited home supply of petrol; therefore, before Japan can mechanicalize her army, she must establish mechanical industries within the country, and assure her petrol supply, either by command of the sea or storage on land. We thus see that maintenance is the link between fighting force and national power; consequently the structure of the military instrument does not only depend on the nature of the resistance it may meet, but also on the resources of the country it is protecting. Maintenance, to be reliable, must be based, therefore, on a correlation between military demand and national supply.

Similarly with repair, if the military instrument is so designed that its repairs demand such highly-skilled labour that the fighting forces themselves cannot provide it, or a sufficiency of it, unless the nation can do so without detriment to itself, the military instrument will either fall into ruin or actually injure the nation it is intended to protect.

Just as we obtain certain relationships between supply and structure, and repair and structure, so do we obtain others between transportation and structure. We know that the fighting and administrative services have to move, but though we realize that these movements must be synchronized, we consistently fail to appreciate the fact that, whilst but a few years ago movement on land was based on muscle-power, to-day maintenance is largely based on mechanical-power, and fighting force on muscular. We still find infantry considered the decisive arm, an arm with a maximum speed of three miles the hour, and with a radius of action of less than twenty miles a day over a continuous period. In the past, the supply and baggage columns of an army were called its impedimenta, because they delayed its progress on the line of march. Now it is the reverse, and so complete is this *volte face* that when infantry require to move rapidly they empty their lorries and get into them. The most efficient relationship between the combatant arms and the administrative services is one which is based on a common means and speed of movement; because similarity of means and speed simplify structure and maintenance, and consequently facilitate control.

## 5. The Control of the Instrument

The military instrument is the weapon of the commander; it is his body through which his will manifests and attains expression; and as a very intimate relationship exists between the brain and the body of man, so should an equally intimate relationship exist between a commander and his command.

With nations such as ourselves we find that the military instrument comprises three great Services—an army, a navy, and an air force; and if these are not controlled by one brain, unity of action, and, consequently, economy of force, are not possible. If these three Services are so organized that it is beyond the powers of one man to control them, the defect must lie in their structure, for, if we accept the human body as our model, control is always possible. We cannot dispense with control, and we can change structure, since the powers of each of the three Services are compounded from identical elements. To hand over war operations to three separate controllers is tantamount to giving a man three heads. When this is done, a monster is created, and, be it remembered, that Cerberus fell victim to the first man who used his one head against its three.

From the mythological aspect of control I will turn to history, and what do we see? We discover that the greatest commanders the world has seen have been those who possessed the fullest powers of control over the instrument of war, and, consequently, over the military instrument, whether it consisted of one or more than one Service. Alexander is an autocrat, for he commanded not only the civil instrument but also the military, and his military instrument comprised both an army and a fleet. Hannibal's failure is due to his lack of control over the civil instrument. Cæsar's success lies in his power to control it. Gustavus is king and general; Marlborough is a generalissimo—he commands on land and sea and, through his wife, he controls the Government at home. Frederick is an autocrat and so is Napoleon. My object here is not to accentuate the desirability of autocracy, but that, if, in war, control is essential, then the freer the will of the commander the more economical will be the expenditure of force.

If we again turn to history, there can be little doubt that many of the great captains of the second degree were in genius equal to these autocrats, but because they were not autocrats, they were unable to attain an equal share of fame. The one power they lacked was complete unity of command, and the more they were restricted in asserting this power, the less were they able to make use of their genius to direct even the purely military resources at their disposal towards gaining their object.

Unity of command expresses unity of will, and, as in the human body, military unity of will and of purpose ultimately find expression in the will of one man. Napoleon understood this full well when he said: (in war) "men are nothing; it is one man who matters"; and again: "The secret in war does not lie in the legs; it resides entirely in the brain that sets the legs in motion." Not the brain of the soldier, but the brain of the general-in-chief. Machiavelli, no mean judge of war, was equally emphatic; he said: "Let only one command in war: several minds weaken an army."

I have laboured this point, because the supremely important fact to be deduced is that, as the object of war is one, control is one, and if this control is shared between several, then the objective cannot be economically gained. In the last great war this veritable axle-pin of generalship was removed from the chariot of command. For four years the Allied Armies floundered through what I believe history will one day denote as a series of the most uneconomical campaigns ever fought in a war of the first magnitude, and, only after a stupendous squandering of lives, resources and money, was the axle-pin pushed home and the war won.

If power of control, vested in one man, is essential, equally is it essential that the structure of the military instrument should be such that it will react to this control. Alexander possessed genius and control, but had he been given the hordes of Darius in place of his superb little army, it is most unlikely that he would have conquered the known world of his period. The military instrument must, therefore, be so fashioned that it can be controlled. In structure it must be simple, its maintenance must be easy, and its whole organization must work automatically, so that the will of the commander can be concentrated on the expenditure of its force.

When I say that power of control must be vested in one man, I mean this in the fullest sense of the words, but I do not mean that one man constitutes the machinery of management.

To revert to an army; besides its commander, it possesses a headquarters which, like the human brain, is " a great administrative governing machine." A portion of the brain (particularly the grey matter in the medulla oblongata at the base of the brain) and spinal cord regulate the reflex activities of the body " without any voluntary control, or even without any consciousness, on the part of the individual "[1]; the directing portions are free to control volition. A similar division of work should be established in every headquarters, management being separated from command, so that command, which eventually must be centralized in the brain of the commander, is free from all routine duties. Thus freed, the brain " can not only drive machines; it can invent and create them . . . It balances and determines the fates of armies, fleets and nations."[2]

The brain depends for its information on the senses, and, for the execution of its orders, on the nerves. We thus obtain three requirements to control: information, decision and communication, the third being the co-operative link between the first and second and the expenditure of fighting force.

If information is regarded as the stable base, then the headquarters of an army is the great receiving, registering and interpreting station, the active laboratory of sensations, of thoughts

[1] *The Physiology of Mind* (1877), Henry Maudsley, p. 136.
[2] *The Engines of the Human Body*, Arthur Keith, p. 235.

and of ideas. The system of communication being the link which connects the organs of information to those of management and command. The organization of military control is the same as in the human body, and, when this is realized, to improve existing organization we must study the body of man—the brain and sensory and nervous organs, and attempt to amend our present system of control accordingly.

### 6. The Higher Control of the Instrument

Thus far I have dealt with control in general terms, and mainly with reference to only one fighting force—the army. I have laid down as an axiom that economy in control can only be attained if one man directs the instrument, not only as a military but as a national weapon, and I have quoted Alexander, Napoleon and others. These men were autocrats and dictators, and though even a democratic nation, when reduced to the last extreme by the pressure of war, will appoint such a man to direct its course, it is too much to expect a democracy to agree to dictatorship, either during peace-time or at the beginning of a war. Though democratic government is government by mediocrity, it is useless kicking against these pricks, therefore it is useless suggesting autocratic control of the instrument, for this would necessitate the selection of a genius as the controller, and nothing a democracy hates and fears more than genius; to the democrat genius is a Satanic force.

In chapter iii. I examined the threefold order of national power, and in this present chapter I have explained that the nation itself is the instrument of war: the question now arises, how can we establish a workable piece of machinery which will control the national forces without infringing the principles of democracy.

Of these principles, the underlying one is rule by the will of the majority, and, as this will is always fluid and consequently always changing, it is not possible to expect careful and progressive war preparation on the part of any democratic government. The masses do not like war, for they are cowardly; therefore their political representatives shun its preparation.

We cannot do away with democratic government, but we could, I think, establish within a democratic nation an advisory council which would consider the question of national defence, which would arrive at decisions on this question and place these before the government for their consideration. In an empire this council would be imperial instead of national.

The organization of this council should follow on the lines of the threefold order. Under it should be established three great departments:

(i.) A department of national (or imperial) ethics, to study national psychology, legislation, local opinion, education and propaganda.
(ii.) A department of national (or imperial) economics, to study national resources, finance, tariffs, science, industry, agriculture, commerce and emigration.
(iii.) A department of national (or imperial) defence, to study the grand strategy of the nation.

These three departments would furnish the council with all possible information for correlation and consideration.

Once having co-ordinated the national powers which go to build up the national instrument of war, the next step is to co-ordinate the fighting Services so that their forces may be economized.

The organization which suggests itself, if the threefold order be kept in mind, is one similar to that of the national council, and as this organization must come under the control of the government, I will call it the ministry of national defence. Its functions should be as follows:

(i.) Ethical: To establish harmony between the three Services and between the Services and the nation.
(ii.) Economic: To divide the bulk sum, voted yearly by the Government for purposes of defence, among the three Services proportionately according to policy and to assess the resources of the country for war.
(iii.) Defence: To convert the policy of the national council as accepted by the Government into a combined plan of action.

We thus obtain a threefold order of control within the national, or imperial, body.

(i.) The national, or imperial council, is the soul of the body; it collects the innumerable national and international sensations and reduces them to harmful and beneficial sentiments.
(ii.) The Government is the mind of the body; it receives the sentiments of the national council, and, reasoning them out, decides what is true or erroneous.
(iii.) The ministry of national defence is the muscles of the body; its duty is action, constructive or destructive.

Though this ministry may be directed by a politician, its true business head should be a generalissimo controlling the three Services. Thus, he will direct three instruments as one instrument, and complete control is established.

### 7. The Study of the Instrument of War

I have now very briefly analysed what I mean by fighting force. I have taken the human body as my model, and then, turning to the nation which is a collection of human bodies, I have assembled all the national powers and resources in one group and have called this group the instrument of war ; needless to say, it is also the instrument of peace. Finally, I have ended with one man who, the closer he can control the forces of this group the more economically will these forces be expended.

Now to apply this knowledge. If our intention is to study military history or to work out a military plan, the first thing we should do is to examine the opposing instruments. Two nations confront each other ; what is the degree of fighting force each nation can apply ? In general terms, the answer to this question is a threefold one, namely, the thinking power, the staying power, and the fighting power of the nation and of its military instrument.

What is the quality of its thinking power? Especially what is the quality of the thoughts engendered by its military brain ? If we can discover what type of mentality we are confronted by and we analyse it, we shall be able to discover its strong and weak points, and shall then obtain a clue as how to direct our own will against it. If the instrument is controlled by one man, soldier or politician, then we should analyse his mental characteristics ; if by a group of men, then we should discover the predominating will in this group, for when war breaks out, in all probability this will will exert itself. We must examine the headquarter organization of the military instrument ; is it controlled by one organ or three organs, and, if by three, which is the predominating partner ? For this partner will exert the greatest strategical influence. We must examine the headquarters of each Service ; are they so constructed as to gain rapid information, give rapid decisions, and obtain rapid communication between body and brain, and brain and body ? All these points are points of vital importance to us as a commander, and when we study military history let us be the commander of one or both sides.

Once we have evalued the thinking power of the opposing forces, I suggest that we turn to their staying power and examine all possible questions of maintenance, under the headings of supply, repair and transportation. I suggest this course because I am convinced that strategy and tactics are founded on administration, and that the maintenance of the military instrument is founded on the resources of the nation, not only military, but ethical, economic and political as well.

Staying power is the base of fighting power, and it is fighting power which renders thinking power concrete and objective in war. The structure of the military instrument must enable the highest fighting power to be developed, and if our examination shows us that this fighting power is defective, then we may conclude that thinking power is also at fault; for fighting power expresses thinking power, consequently it is correlated to it.

Fighting power is a compound of stability, activity, and mobility, or of resistance, pressure, and the co-operative energy engendered by these two. The protective, close combat, and pursuit troops of an army are its two arms and its legs. What are their individual values and their combined value? If we can discover these we shall understand their tactical values, and, in history, we shall be able to watch how they have been used; or, on active service, understand how to use them.

To conclude: in war we are faced by a nation, which is the instrument of war we have to meet. This nation possesses a civil and a military side, and the correlation between these two sides is grand strategy. The civil side is the base of the military side. The civil side comprises ethical, economic, and political power, all of which are means of war. The military side—an army, a navy, and an air force, or at least one of these forces. The military side is built out of three elements, and these three elements govern the structure, maintenance, and control of the military instrument. In an army, we must have three types of troops, namely, protective, close combat, and pursuit troops; we must have three systems of maintenance, supply, repair, and transportation; we must have three means of control—information, decision, and communication. Here are nine factors which give character to fighting force. What is its value? This question I will attempt to answer in the following three chapters.

# CHAPTER VI

## THE MENTAL SPHERE OF WAR

He who will not reason is a bigot, he who cannot reason is a fool, and he who dares not reason is a slave.—SIR W. DRUMMOND.

The beginning of all Wisdom is to look fixedly on Clothes . . . till they become *transparent*.—T. CARLYLE.

### 1. THE ELEMENT OF REASON

IN chapter iii. I examined "The Threefold Nature of Man," and I showed that it comprised three spheres of force—the mental, moral, and physical. In this and the next two chapters I will consider these, and in the present one the first.

As the brain and the nervous system control the body, and as the national head (King or President) and his Government control the nation, so also does a general and his staff control his army, or a generalissimo and his staff the combined fighting forces placed under him. In each case the aim or purpose is the same, the means alone change, and there can be no doubt that, if in the last two cases the control were as complete as in the first, both a nation or its military forces would become amazingly efficient instruments. I intend, therefore, to open this chapter with a brief examination of the controlling faculties of the mental sphere, namely, the reason, the imagination, and the will.

When I speak of mind, I am thinking of the intellectual qualities of man, of his thoughts, his ideas, and the decisions he arrives at. Man is a conscious animal; whatever he perceives is the result of sensation; all his experiences are based on sensations, and all his knowledge is ultimately based on experience. Though the data of experience are divided into several states of consciousness, in all of these we can discover three elements, namely, feeling, the forms of feeling, and the remembrance of feeling. The feeling itself may be compared to a plastic substance upon which is imprinted every sensation which is conveyed to it by the senses. The second are the categories of sensations, and these depend on the senses themselves; thus, there are categories of sight sensations, of hearing, of touch, etc. The third endows feeling with a power to recognize two or more sensations of a similar nature, the new ones awakening the old.

Sensations are the only facts vouched us to work on, for they form the material of the mind, they give birth to thoughts, to ideas, and, finally, to judgments.

In the objective world errors do not exist; all things are controlled by law which works automatically and not consciously. Errors are subjective, they are the privilege of the mind, and so also is truth, which is not Reality but its reflection. We thus obtain two moods of reason, one which correctly reflects Reality and the other which contorts the reflection. We cannot abolish error and, if we could, we should possess no standard whereby to judge truth. It is through error that we arrive at truth, but only if we can rationally discover the degree of error. This means that we must understand our errors: what is their cause; what is their effect; whence do they come; whither do they lead? To answer these questions, we must understand the reasons for error. It is not that error excludes truth, or truth error, for they are moods of reason, and are consequently inseparable. Error is our teacher and truth the marks he allots to us for good work, and good work is accomplished by correct thinking, which is arrived at by less and less erroneous thinking.

What has all this got to do with war? Everything! There must be a reason more or less erroneous or true for a war, otherwise the war is a struggle of maniacs. There must be a reason for each action carried out during a war, and again it must be a good reason or a bad reason; and if we have no reason at all, which has frequently happened in war, we reduce ourselves to the position of lunatics.

If we understand the true reason for any single event, then we shall be able to work out the chain of cause and effect and, if we can do this, we shall foresee events and so be in a position to prepare ourselves to meet them. Our reason is the director of our actions and also the spirit of our plan. If we fail in our purpose, in place of blaming circumstances we should blame our reason, for the main fault lies there. We must analyse its motive and discover where it has failed us; thus, we shall turn errors to our advantage by compelling them to teach us. We must not allow ourselves to be enslaved by them, for they should be our masters, not our taskmasters.

Reason is the highest form of consciousness, it draws its "substance" from memory and, in the light of the imagination, it focuses memories according to the conditions of the moment. In war, as in peace, reason is the controlling faculty of the mental sphere. All our conscious actions emanate from reason, just as all our bodily activities emanate from physical force, and, as I shall explain in another chapter, because military power is controlled by similar laws to those which govern force, consequently the one aim of the soldier is to harmonize his mind to the workings of these laws.

### 2. The Element of Imagination

If war were an exact science, reason in itself would be all but sufficient to arrive at correct judgments, but it is far from being exact, since it deals with the differences between living creatures in place of inanimate substances or quantities. In mathematics, two multiplied by two is always four, and in chemistry two molecules of hydrogen and one of oxygen always form water; but in psychology, and war is largely a matter of psychology, two ideas in one man's head do not necessarily lead to the same judgment as two similar ideas in another man's head, because each individual possesses a faculty called imagination, and no two imaginations are constant.

In war we deal, therefore, not only with known quantities—the organization of the enemy's army, its strength and equipment, and the nature of the theatre of war, concerning which reason is our paramount guide—but also with a host of unknown or partially known quantities and qualities, the larger proportion of which are psychological in nature, and concerning which we must work by means of hypothesis.

I have already examined the value of hypothesis in chapter ii. If in the civil sciences it can help us, how much more so can it assist us in the science of war.

Some men are born with an all-illuminating imagination, but these men are few in number. The average man possesses little or no imagination; how then can he cultivate it? We cannot endow him with a natural faculty, if this is wanting, but we can supply him with a synthetic substitute, which will partially make good the deficiency. We can show him what history has to relate concerning various operations, situations, and things. If certain results have occurred again and again, and it is discovered that certain factors and circumstances have been common when these results were obtained, then we may infer the likelihood of similar factors and circumstances producing like results. The man of imagination would see the results spontaneously, for as I have said, his imagination would focus his powers of reason and lead him directly to this deduction.

Take another case. A little imagination will lead us to realize the difference between our mentality and that of a Frenchman or a

German; and once we have realized this difference, we can instantaneously assume the mental attitude of a Frenchman or a German, and see things as they would see them: this is a most important factor in war, this stepping, not into our adversaries' or friends' shoes, but into their minds. Few men, however, can do this, but once again a careful study of national characteristics will enable them approximately to obtain a foreign point of view, and to understand the psychology of their friends and foes. If a general knows that the racial characteristics of his enemy are *a*, *b*, and *c*, and the individual characteristics of the opposing general *x*, *y*, and *z*, then he will be able to act accordingly. This knowledge gives him an immense advantage. If besides this knowledge, he possesses so acute an imagination that he is able to sense the moral, rather than mental, worth of his antagonist in his actions, then his advantage is immeasurably increased. He, in fact, possesses what is called genius, a quality I will examine a little later on in this chapter.

### 3. The Element of Will

In the second chapter of this book I stated that, if thoughts are fixed in one direction by a conscious impulse, the result is will. Will is, so to speak, the gravity of the mind,[1] it is the motive force which attempts to accomplish reason by cause and effect. Thus, to make a comparison: a stone thrown up into the air eventually gravitates towards the centre of the earth, but only reaches the surface, since the force of gravity is not equal to the resistance the earth offers to its progress. If we could sufficiently reduce this resistance, or increase the force of gravity, the stone would be pulled through the earth and eventually reach the centre. As the aim of gravity is to bring the stone to rest at the centre of the earth, where all activity ceases, so in war the aim of a commander's will is to bring his enemy to rest; in fact, to deprive him of all power of movement. To do so he must either reduce the resistance the enemy is offering to his will, or increase the powers expressing his will to so high a degree that his own will can move as gravity moves the stone along the shortest path between his reason and his goal. In the first case, he must compel the enemy to distribute or disperse his resistance, and, in the second, he must concentrate his force, his will, and its means of expression; and the more he can force the enemy

[1] "Will is not an entirely unknown quantity; it indicates what it will be to-morrow by what it is to-day . . . each of the two opponents can . . . form an opinion of the other, in a great measure, from what he is and what he does," instead of what he should be and should do. *On War*, Clausewitz, vol. i., pp. 7, 8.

to disperse his strength, and the more he can concentrate his own, the more direct will cause, if it be well founded on reason, produce the required effect.

Though the desired aim in war is to impose one's own will on the enemy, the two wills in conflict are surrounded by a host of other forces. Thus, each will depend on the reason of the action contemplated; each on how far this reason is free from error. Again the will of each commander must find expression through the will, individual and collective, of his men, and, in turn, their will depends on how far they can subordinate it to his, and how far their means of expressing it are or are not superior to the enemy's.

It is easy enough to say that the aim of war is the imposition of one will on another; but for a moment examine this statement and it will be seen how complex it really is.

First, each of the opposing wills is attempting to express a reason in order to gain an end. Which reason is the soundest; which brain has evolved the better plan of action? Which side has foreseen how its plan will shape itself, and which side is prepared to modify its plan without abandoning its motive?

Secondly, which side has more effectively attuned the wills of its men to the will of their commander. Which side possesses the highest self-sacrifice, the staunchest discipline, the firmest loyalty and closest comradeship? Then, when the will of the commander can no longer direct, which side will substitute a collective impulse for his individual impulse, and control the course of action as if their commander were standing behind them personally directing events? As an architect plans a house and as the masons build it, so must the plan of the commander be executed by his men in detail, whether he be near them or far away. Here again it is the plan which is the guiding and directing force, and its execution depends on skill and will to carry it out.

Thirdly, will demands means of expression. Are our means superior to those of the enemy? Skill is not sufficient; for deprive the skilful worker of his tools and his talent and ability are at a discount. If he feels that he is out-tooled and cannot move as the enemy moves, hit as the enemy hits, and protect himself as the enemy protects himself, his *moral* will fall, and, as it falls, so will fear jostle aside his endurance, obliterate or unhinge his will, and cut it off from that essential co-operation with the will of his commander, and so reduce a rational plan to an irrational struggle.

The imposition of our will on the enemy may be the whole aim of war, but will is an element attracted and repelled by the other elements; consequently we must understand what attracts it and what repels it, what accelerates and retards its activities, for not until we understand these things shall we know how to impose our will and how to prevent the enemy imposing his will on us. The imposition of will is the statement of a fact; how to impose it is, to the normal man, a lifelong study of the elements of war and of their relationships.

### 4. The Influence of Genius

If we now turn to the history of war we shall soon discover that, in every period in which the art of war has progressed rapidly, the cause of this progress is the mind of some one man—an Alexander, a Hannibal, a Gustavus, or a Napoleon. To us these great captains appear to possess a natural gift for doing what is right and shunning what is wrong, and this gift is called genius.

Genius is one of those apparently inexplicable powers which differentiates the truly great man from the normal. It is not an instinct, for otherwise it would be common property; it is not reason, as we usually understand it; but, as it accomplishes in an incredibly short time a purpose which the faculty of reason would attain by a slow and no more certain progress, it, I think, may be considered as the highest dimension of this faculty. Whilst the mass of mankind shows little reasoning-power and relies on imitation—the crowd instinct—the man of genius transcends mere copying; he refuses to swim with the stream; he strikes out in a direction of his own; and, what appears almost a miracle to the crowd, he frequently succeeds in diverting the stream from its course by compelling it to swirl forward in his own direction.

The military genius[1] is he who can produce original combinations out of the forces of war; he is the man who can take all these forces and so attune them to the conditions which confront him that he can produce startling and, frequently, incomprehensible results. As an animal cannot explain the instincts which control it, neither can a man of genius explain the powers which control him. He acts on the spur of the moment, and he acts

[1] Lloyd says of the military genius: "Great geniuses have a sort of intuitive knowledge; they see at once the causes, and its effect, with the different combinations, which unite them: they do not proceed by common rules, successively from one idea to another, by slow and languid steps, no: the *Whole*, with all its circumstances and various combinations, is like a picture, all together present to their mind; these want no geometry: but an age produces few of this kind of men: and in the common run of generals, geometry, and experience, will help them to avoid gross errors" (*History of the Late War in Germany* (1766), Preface to vol. i., p. 19).

rightly, because this power is in control. That some explanation exists cannot be doubted, but so far science has not revealed it, though the psychologist is working towards its fringe.

When we look over the history of war we see no steady growth; in place we see revolutions in the art, and fallow periods. These revolutions are rapid and short, for they invariably coincide with the life of some genius. In the art of war Alexander accomplished in twelve years more than had been accomplished in the twelve thousand years which preceded him. His work was not all his own. He borrowed from his father, from Xenophon, from Cyrus, and others; but his genius compelled him to borrow what was right, and it repelled him from copying what was futile.

How is it that such geniuses flame over the horizon of war like shooting stars, scintillate for a little, and are gone, and fallowness so frequently follows in their path? One reason is that genius is a rare quality of mind, and it is unusual that one great man is followed by an equal, and another is that, until we possess a true science of war we have no means of calculating the results of genius. An Alexander comes, he conquers, and he goes, and, though thousands have watched and followed him, to them his genius remains a mystery. The man is venerated, but his method vanishes, not because it is forgotten, but because it was *never* understood.

If military genius possesses the power of producing original combinations from the forces of war, genius must consequently be the mainspring of strategy, which is largely the science of forces. Inwardly its work is founded on originality; outwardly it manifests in surprise. The great genius surges through difficulties immune, because he sees—foresees—the end, and understands the means. It is his mind which tramples down his enemies, though seemingly the weapons of his men accomplish this end. If *moral* is to the physical as three to one, then genius is to the normal as thirty to one. True, a man of genius may be overwhelmed—some have been—but, to appraise such a man, his worth must be judged not so much by the successes he has gained as by the art he has created. For it is what is endurable in the soldier and his art which constitutes the Golden Fleece of our quest and the reward of our studies.

The first master of the art of war is experience, the second is reason, and the third, and greatest, is genius. Experience can be bought at its price; reason can be obtained by study and by reflection; but genius would appear to be God's gift. In other words, if we cannot understand cause and effect, we must sense their relationships, and so add something to our stock of knowledge. Again, if we can reason out cause and effect we discover their relationships without loss of energy; but it would appear that what the man of genius does is to imagine automatically, and so produce original relationships which, metaphorically, are born patented, since others can seldom copy them.

If I may hazard to set down the qualifications of the great captain, then I should say that they are:

(i.) Imagination operating through reason.
(ii.) Reason operating through audacity.
(iii.) And audacity operating through rapidity of movement.

The first creates unsuspected forms of thought; the second establishes original forms of action; and the third impels the human means at the disposal of the commander to accomplish his purpose with the force and rapidity of a thunderbolt. From the mind, through the soul, we thus gain our ends by means of the body.

### 5. The Vehicle of Genius

As genius is a personal gift, so is imitation a collective instinct. One man possessed by genius may alter the course of history, in fact, such a man has always altered the course of history, when alteration has been rapid. Three men of genius, working as a committee, could not do this, and still less so a crowd of normal men.

Whether genius can actually be cultivated or not, I cannot say. I have suggested that a synthetic genius[1] can be cultivated, but a more important question is: Can we train our minds to recognize genius? I believe we can; if I am right, then when a genius appears we shall not impede him, for, if we can recognize him, we shall be able to assist him. Here our predominant difficulty is the spirit of the herd, which in these democratic times has been deified and raised to Olympian heights. As long as the herd-spirit controls a nation, men of genius may be born, but circumstances will prevent them spreading their wings. Only picture to ourselves a supreme financial genius entering the department of the Treasury! What could he do? He could do no more than George Stephenson could have done had he suddenly materialized in the camp of Boadicea. Genius, for its expression, demands, therefore, conditions in which it can express itself; this is what we must realize, and especially so

[1] Synthetic genius attains its end by cultivating aptitude in the correct application of the principles of war.

when we deal with war. We, as pioneers, must blaze the trail for genius; we must cease relying on traditions which in their day may have been excellent, but which in our day are threadbare.

What does this preparation demand? It demands clear thinking.

Since we cannot breed men of genius at will, this is then our problem: to think clearly; and what is the first step in its solution? To cease imitating. I have already pointed out the short-sightedness which characterized the period immediately preceding the outbreak of the Great War of 1914–18. In spite of this war, this period is not dead; in fact, it is very much alive, for whenever anything new is suggested we are urged to proceed with caution, ever forgetting that fear is failure and the forerunner of failure.

Caution may be an excellent precept, but none the less so is audacity, yet what is still more excellent is to think clearly, for clear thought leads to true thought, and, once a truth is grasped, the sooner we make use of it the better; for, if it be a truth, then as long as we do not full-heartedly accept it and mould our opinions and actions upon it we shall simply be maintaining and fostering a lie.

Why is caution always on our lips? Because we are not sure of ourselves, because we openly, or hiddenly, acknowledge our ignorance. As long as we are ignorant this is excellent, but do not let us make caution an excuse for remaining ignorant—do not let us canonize it. It is very easy to do so, and sometimes, to the mentally inert, it is very comforting to have a saint. Instead, let us say to ourselves: I am proceeding cautiously because I am ignorant; I must overcome my ignorance so as to step out audaciously. Clear and valiant thought—this is our sword.

Another frequent excuse for remaining indolent is the expense entailed in effecting a change in armament, or equipment, etc., yet it cannot be doubted that an obsolete army is the most expensive organization a nation can maintain, since it cannot fulfil the purpose for which it is established, namely, to secure the nation against war, or, when war comes, to terminate it rapidly. Sometimes this excuse is openly based on indolence, but more frequently because anything new is apt to upset vested interests. Traditionalism is a herd-force, and vested interests are armoured with traditions; so much so is this the case that mobs and mob-rule, throughout history, have remained psychologically unchanged.

Change, to be really productive, must be systematic and objective. It must be attuned to needs and not to fancies. It is not sufficient to invent something novel, but something useful, and to do so we must fix the end before we change the means.

To take weapons as an illustration, in the past and to-day how do new weapons appear? Some enthusiast, frequently a civilian, sees a tactical defect, and introduces a new arm to make it good. The soldier quite possibly has not seen the defect, yet the arm is adopted as it may prove useful. It is glued on to the existing organization, and at the first shock it chips off. It is then pronounced useless, when, in fact, it might, if correctly used, prove of the greatest service.

In an army every material novelty demands first a clear, tactical appreciation of its use, and secondly a suitable organization, based on this appreciation, wherein to express its powers. Improvement in means should be based on clear-cut ideas; in fact, tactical demand should precede technical supply. I want a weapon of such a nature because I want to carry out tactics of such a nature, and not, Here is a new weapon; what are its tactics? should be the guiding rule in change.

From these few examples I think the student will see that we cannot sit down and wait for genius to rectify error. In all probability, in no period in history have men of military genius been wanting. What is scarcer than genius is opportunity propitious to its manifestation. In the past, opportunity has frequently been created by some great turmoil, such as a revolution, which, pulverizing traditionalism, has liberated the man. This is a sorry method; surely we can do better than this; surely we can abandon obsolesence without disintegrating a whole nation; surely, knowing as we do that we possess a faculty called reason, we can prepare the way. How to think rationally, this is the problem I have set myself to solve, and not how to endow the student with genius, for, in my opinion, reason is the first element of war, from which the directing force of all the other elements emanates.

### 6. Military Thought

The process of rational thought is the same for all men, and this process I have already explained in my lecture on the method of science. The process must be applied to some definite end, and our end is war.

Though the art of thinking is a very ancient art, and though logic has controlled philosophy and science for hundreds of years, logical thought has not been applied to war, except by a very few; because logical thinking demands the arrangement and organization of thought according to the values of the subjects of thought and the objects these subjects represent, and, so far, method has been wanting.

In war—perhaps more so than in most other activities—a good reason is not necessarily a true reason. Knowledge and understanding possess immense force, yet unless they are correlated by wisdom their very power may prove a danger. A wise man is not only a man who knows, but a man who sees and knows; he is, in fact, a man of common sense, a man who possesses the power of adapting thought and action to circumstances, and to do so he must understand the circumstances.

A wise soldier is like a wise surgeon; he is faced by an operation, but, possessing skill and knowing intimately the anatomy of war, he can operate judiciously.

And what is the anatomy of war? It is much the same as the anatomy of the human body, since armies are human organizations. In war, armies face armies; they possess structure, control, and maintenance; their forces are developed in three spheres—the mental, moral, and physical—and are expended in varying circumstances. Here we have three things we must consider—organization, force, and circumstances—and it is wisdom which sets these three in harmony.

Knowing much, and seeing the changing conditions which surround him, the skilled soldier will always be seeking for new ideas whereon to mould his plan. An idea strikes him; it surges out of his memory, awakened by some sudden event. His first step is not to apply it, but to mould it; and it is this process of shaping ideas into practical plans which is so difficult, unless the soldier possesses genius or method.

The first thing to remember is that a new idea should not necessitate a sudden change in structure. Structure can of course be changed, but only slowly, and, in war, if it be rapidly changed, the control and maintenance of an army may be detrimentally affected. Generally speaking, novelties must be limited to work within the existing organization; in other words, a brilliant idea will prove even dangerous unless it can be applied without necessitating a rapid and radical structural change.

Remember also that in battle, and battles are the tests of military structure, the object of each side is not to kill for the sake of killing, but for the sake of disorganizing, for military strength does not reside in individuals, but in the co-operation of individuals and masses. Co-operation depends on control; and the endurance of force depends on maintenance. Every plan must have a threefold base; it must permit of the existing structure of an army remaining unaltered, or as unaltered as possible; it must permit of the existing system of control working without friction ; and it must permit of the administrative units carrying out their duties without let or hindrance.

If the student agrees with what I have now said, before he attempts to transmute an idea into a plan of action, he will carefully consider the influence of his idea on the general organization of the force he intends to apply. He will consider how it will affect tactical organization, the organization of command, and the organization of administration, and, having decided on the answer, then he can consider the second point.

Organization is the vehicle of force; and force is threefold in nature; it is mental, moral, and physical. How will the idea affect these spheres of force? This is primarily a question of force and its expenditure. Thus, if the idea is complex, and does not permit of it being readily grasped by others, mistakes are likely to occur; and if its aim is beyond the moral and physical powers of the troops, should it be pushed beyond the limit of their endurance, though organization may for the time being be maintained, ultimately demoralization will set in, and a demoralized organization is one which has become so fragile that a slight blow, especially a surprise blow, will instantaneously shatter it to pieces.

The third point is that the idea must not only harmonize with existing conditions, but with their probable fluctuations. This is a most difficult factor to gauge, and it is here that the man of genius transcends the normal commander. Failing genius, it is by imagination that we can overcome this difficulty. Every action will produce a definite effect; and if we are not endowed with imagination, then we must fall back on reflection, and work out mathematically the chain of cause and effect, not only from our own standpoint, but from that of the enemy as well. Thus: my idea is A, and existing conditions are B; my first move is X; what will the enemy's be? It may be Y or Z. How will Y or Z affect B? Y may not alter B, but Z may produce a new series of onditions—B + C. What, now, will be the influence of B + C on A—and so on?

We first look at the idea or plan from our own point of view, and then from the enemy's, and discover, not only what these two points of view are, but how they will influence existing conditions, and how these conditions will change.

These, in brief, are, I think, the most important points in applying military thought to a problem: maintain organization, work within the limitations of the force at our disposal, and foresee the changes in conditions.

### 7. Grand Strategy

In chapter v. I outlined the machinery of control, and in this present chapter I have examined the force which this machine should liberate. I have explained how the brain in part works automatically and part consciously; and it is the same in war, for what is required is that the duties of peace, which must continue, should work automatically, so that the government may concentrate the whole of its attention on the war and render every fact concerning it a conscious and a considered fact.

In war, as I have explained, a government works directly through its own political weapons, and indirectly through its military instrument. Thus in war a government is concerned with three great duties; namely, to maintain the domestic machinery of the nation; to set in motion the political machinery; and to control the military machinery. The first is the base of the second, and the second of the third, and all three must be correlated.

I have shown that economy demands that the fighting forces should be directed by a generalissimo, and by a generalissimo I do not mean a *fighting* commander-in-chief, but a thinking man, assisted by a highly trained staff drawn from the three Services. A man who can free the fighting commanders—whether operating singly or unitedly under one chief—of the formulation of policy, and of direct political interference. In most modern wars, and conspicuously so in the Great War of 1914–18, each commander-in-chief had to face two fronts—the enemy and his government; the result was that pressure in rear hindered command in front. Throughout the last war the appointment of commander-in-chief was purely nominal; no such officer really planned, really commanded, and really fought, for command was by delegation. It was a war which Gustavus, Frederick, or Napoleon could not have dreamt of.

What is now required is a system which will liberate the fighting head; and, as democratic nations will not tolerate the appointment of a military dictator, unless they are on the point of being deafened by their death-rattle, the only remedy would appear to be to establish a military buffer between the government and its instrument.

The generalissimo should be, therefore, the thinking, co-ordinating head, who can advise his government on the formulation of the grand strategy of the war, which, in the main, is the correlation between national power and military effort; for grand strategy includes all the forces which are to be expended in the struggle. "No war," writes Clausewitz, "is commenced, or, at least, no war should be commenced, if people acted wisely, without first seeking a reply to the question, what is to be attained by and in the same? The first is the final object; the other is the intermediate aim. By this chief consideration the whole course of the war is prescribed, the extent of the means and the measure of energy are determined; its influence manifests itself down to the smallest organ of action."[1]

This is grand strategy. How, then, can a commander-in-chief (unless he be a dictator) concentrate the whole of his mental energy on the prosecution of the war unless he is freed from political interference. If, on every occasion upon which he wishes to do anything, he is compelled to refer the question to a many-headed cabinet, the members of which possess no strategical knowledge, opportunity will vanish long before decision is reached. If, on the other hand, he is able to refer it to a generalissimo, whose duty it is to keep in the closest touch with political affairs, he will be told forthwith whether his actions coincide or run counter to policy.

I have in a former chapter examined the forces which build up national power, and in another, the object of war in its threefold order. It is these that the grand strategist has to correlate with the conditions of war actual and problematical, so that the force of the instrument of war may be expended at the highest profit. It is for this reason that in the last chapter I have suggested that his department should be organized to deal with economic and ethical questions as well as defence. His office should work in closest co-operation with the national council, so that between these two the political mind of the nation will not only be equilibrated by this dual pressure, but brought into the closest touch with the realities of war and the realities of national life as influenced by war. Without some such mental pressure policy must remain inarticulate; the politician, on the one side, fearing public opinion, and, on the other, distrusting the will of the army. Without stability of policy there can be no stability of plan, and without stability of plan there can be no economical direction of force.

Whilst in the past, when nations were more self-contained and less interdependent, the grand strategist was, normally, a soldier who at times controlled both the land and the sea-forces, and who was endowed with political instinct; for example, such men as Cromwell, Marlborough, and Napoleon; to-day the grand strategist must be something more than these great men. He must be also a psychologist and an economist; and, as we

[1] *On War*, vol. iii., p. 79.

can never guarantee that when war is declared we shall find a genius in control, we must create so perfect a piece of grand-strategical machinery that a man of normal intelligence and high training will be able to carry out the duties of grand strategy with effect. Failing genius, it is the machine which will produce the man, not a fighting soldier, sailor, or airman, or these three combined in one, not a fighting head, but a thinking head, a centre of thought—a war brain, which will direct the forces, but not the activities of the instrument.

### 8. Grand Tactics

The correlation of the forces of war is the main duty of the grand strategist, and, once these forces have been correlated and adjusted to the political object, the next step is to endow them with structure so that they can be operated. This is the duty of the grand tactician; he takes over the forces as they are distributed and arranges them according to the resistance they are likely to meet. This arrangement constitutes the plan of the war, or campaign, and, if the spirit of the plan is the political object, then the heart of the plan is the military object. This object I will now consider.

In war the object of military action is to compel the enemy to accept the policy in dispute; it accomplishes this end by disarming the enemy and occupying his country, which renders it possible for the government to impose its will on the hostile nation with honour and economy. Or as Clausewitz says: "There are three principal objects in carrying on war:

"(*a*) To conquer and destroy the enemy's armed forces.
"(*b*) To get possession of the material elements of aggression, and of the other sources of existence of the hostile army.
"(*c*) To gain public opinion."

The first, he says, is gained by defeating the enemy's army; the second by occupying those points at which resources are concentrated; and the third by great victories and the possession of the enemy's capital.[1]

These three objects (though to-day the means of attaining them are somewhat different than they were a century ago) agree very closely with the national, ethical, and economic objects I examined in the last chapter.

As grand strategy secures the political object by directing all war-like resources—moral, physical and material—towards the

[1] *On War*, vol. iii., pp. 209, 210.

winning of a war, grand tactics secures military action by converging all means of waging war towards gaining a decision.

The grand-tactical object is the destruction of the enemy's plan, which destruction will so reduce his will to win that he must either surrender or accept terms of peace. The strength of this plan is, however, divided between the hostile army, government, and people, all of which should, if possible, be attacked directly or indirectly by force of arms and by political action.

When Clausewitz wrote his famous book he only considered the operations of armies which by the nature of their structure are compelled to fight in two dimensions. In his day, and until quite recently, it seldom was possible for one nation to impose its will on another without first destroying the enemy's army, or by gaining so decisive a victory over it that the national will was left unprotected; consequently Clausewitz lays down that: "The overthrow of the enemy is the aim in war; destruction of the hostile forces, the means both in attack and defence."[1] Nevertheless, he realized quite clearly that this overthrow, in its turn, was only a means of enforcing policy; yet most of his followers have glossed over this important point, until in the political and military minds destruction has ceased to be a means and has become an end in itself.

Though Clausewitz saw, I think, clearly the political side of this question, on the military side he seems to have lost his way, and it is for this reason, I imagine, that his students have done likewise.

At the beginning of his work, in book I, he appreciates the fact that "in war it is only by means of a great directing spirit that we can expect the full power latent in the troops to be developed."[2] And a little later on, of the commander, he says: "Ordinary men who follow the suggestion of others become, therefore, generally undecided on the spot; they think that they have found circumstances different from what they had expected, and this view gains strength by their again yielding to the suggestions of others. But even the man who has made his own plans, when he comes to see things with his own eyes, will often think he has done wrong . . . his first conviction will in the end prove true, when the foreground scenery which fate has pushed on to the stage of war, with its accompaniments of terrific objects is drawn aside and the horizon extended. This is one of the great chasms which separate *conception* from *execution*."[3] In fact, this chasm holds, or should hold, the mental endurance of the commander.

In another place Clausewitz points out that the enemy's

[1] *Ibid.*, vol. iii., p. 6. [2] *Ibid.*, vol. i., p. 74. [3] *Ibid.*, vol. i., p. 77.

resistance acts directly upon the combatants, and that through them it reacts upon their commander. "As soon as difficulties arise," he writes, "—and that must always happen when great results are at stake—then things no longer move on of themselves like a well-oiled machine, the machine itself then begins to offer resistance, and to overcome this the commander must have a great force of will. . . . As the forces in one individual after another become prostrated, and can no longer be excited and supported by an effort of his own will, the whole inertia of the war gradually rests its weight on the will of the commander: by the spark in his breast, by the light of his spirit, the spark of purpose, the light of hope, must be kindled afresh in others: in so far only as he is equal to this he stands above the masses and continues to be their master; whenever that influence ceases, and his own spirit is no longer strong enough to revive the spirit of all others, the masses, drawing him down with them, sink into the lower region of animal nature, which shrinks from danger and knows not shame."[1]

The importance of the commander as the vital, mental, and moral centre of his army is wonderfully accentuated by Clausewitz, yet, as he proceeds in the development of his philosophy, he loses sight of this point. In his fifth book he writes: ". . . except the talent of the Commander-in-chief—a thing entirely dependent on chance. . . . The nearer we approach to a state of equality in all these things the more decisive becomes the relation in point of numbers."[2]

Brute force now to a large extent replaces the will of the commander as the vital factor in war, and out of this change, Clausewitz, in part—and I think the greater part—misjudging the art of Napoleon, elaborates his theory of "Absolute Warfare,"[3] which, though to him is "a struggle for life or death," to his followers suggests the idea of "destruction."

I have gone to this length in the examination of this question because our present-day theory of war is based on Clausewitz, possibly on a misinterpretation of Clausewitz, who, I consider, misunderstood Napoleon. To the masses of fighting men, in war, the object of an army is to destroy an army; of a fleet, to destroy a fleet; and of an air force, to destroy an air force; in fact, to these folk, the object in grand tactics is the maximum destruction at the minimum loss, or, more frequently still, at any cost.

[1] *Ibid.*, vol. i., pp. 54, 55, 57. [2] *Ibid.*, vol. ii., p. 6.

[3] *Ibid.* See vol. ii., p. 358 and vol. iii., pp. 79–83. See also my book *The Reformation of War*, chaps. iv. and v.; and Captain B. H. Liddell Hart's analysis of "The Napoleonic Fallacy," in *The Empire Review*, May 1925.

Though in minor tactics this is partially true, in grand tactics I maintain that it is an error of the first magnitude. The decisive point is not the body of the hostile army, just as politically the decisive point is not the body of the hostile nation. Politically, the decisive point is the will of the hostile nation, and grand tactically it is the will of the enemy's commander. To paralyse this will we must attack his plan, which expresses his will—his reasoned decisions. Frequently, to do so, we must attack his troops, but not always; for he can be attacked in rear by the will of his own people and his own politicians, also he can be out-manœuvred and surprised. The grand tactician does not think of physical destruction, but of mental destruction, and, when the mind of the enemy's command can only be attacked through the bodies of his men, then from grand tactics we descend to minor tactics, which, though related, is a different expression of force.

We see, therefore, that grand tactics is the battle between two plans energized by two wills, and not merely the struggle between two or more military forces. Consequently, to be a grand tactician, it is essential to understand the purpose of each part of the military instrument.

### 9. The Purposes of the Fighting Forces

Man is a terrestrial animal, and the only certain method of compelling an enemy to accept the policy in dispute is to occupy his country. Without such occupation it is not possible to guarantee adherence to terms of surrender. As there can be little dispute as to this, I will lay it down as an axiom that the peaceful occupation of the enemy's country is a sure guarantee of success in war; and by peaceful I mean that all armed resistance throughout the enemy's country has ceased

This occupation demands an army, or a police force, that is some form of land-force, which can enforce and maintain tranquillity amongst the enemy's people. If this army is separated from its own country by sea, then to effect this occupation and to maintain it, command of the sea communications leading to the enemy's country is an essential. This in its turn demands a fleet.

From this may be deduced the following: that whilst the object of the army is to create a situation which will compel the enemy to accept the policy in dispute, this situation is only definitely established when the enemy's country has been occupied and all armed resistance has ceased. In other words, the purpose of an army—that is, its *raison d'être*—is to gain command of the enemy's land. Occupation is, in fact, the attainment of this object, for once the enemy's resistance has been overcome the ultimate military objective is won.

As I shall deal with military objectives in another chapter, I will turn to the purposes of a fleet.

It has two:

(i.) To protect the transportation of armies, and to compel the enemy to disperse his main army by landing or threatening to land troops.

(ii.) To protect the transportation of supplies, and to impede or completely prevent supplies being shipped to the enemy's country.

The first is the military purpose of a fleet, and the second its economic purpose, which together may be expressed in one term —command of the sea, or the power of controlling movement over the waters in order to maintain and secure policy.

As the ultimate aim of a fleet is to gain or maintain command of the sea—that is, liberty of movement and action on the water —consequently its object is to clear the sea of all hostile ships, either by sinking or blockading them, and until this objective has been gained the purposes of a fleet cannot without grave risk be accomplished.

Thus far the problem seems clear enough: occupation of the enemy's country is essential; and his resistance may be broken by military pressure, which is physical, or by naval pressure, which is economic,[1] or by both in co-operation.

In recent years this simple problem has been rendered complex by the discovery of flight, and one of the supreme war questions which confront all nations to-day is: how will air-force influence this problem?

Armies and fleets are instruments of political force, which, in order to render this force operative, have, normally, to destroy the enemy's military and naval resistance. An air force can act otherwise; it can, in certain cases, ignore armies and fleets,

[1] The effectiveness of the navy as an economic weapon is little realized by the general public. The following, told me by a naval friend, quoting the highest authorities, is of interest: "Up till the end of 1918 it is calculated that 763,000 German civilians *died* as a result of the 'blockade.' The spread of tuberculosis has undone the work of many years before the war, and a large percentage of the children of Germany are more or less affected with rickets. The new generation will be permanently injured, both mentally and physically. The result of the 'blockade' in terms of human misery was unutterably dreadful, but as a measure of war it can only be described as a wonderful success." It appears somewhat cynical that the economic blockade should be the means whereby the League of Nations proposes to enforce its will.

and directly attack the will of the hostile nation. Possibly, in the future, aircraft may become so powerful that surface fleets and armies will be unable to protect themselves against them. In the first case, the older forces are ignored, and in the second they are destroyed, and if the terror wrought by aircraft is so great as utterly to paralyse a nation, occupation may be effected by merely walking over the frontiers.

I do not say that this is an impossible eventuality, but, remembering the limitations which landing-grounds and gravity impose on aircraft, I am of opinion that, until a new motive power is discovered and aircraft are radically changed, the true purposes of an air force are :

(i.) To provide the army and navy with information and local protection.
(ii.) To attack the will of the hostile people.

The first is the military and naval purpose, and the second the moral, or psychological, purpose, both of which are gained through command of the air.

As all three Services—army, navy and air force—are based on the land, the army, in its turn, must co-operate with the navy and air force by protecting these bases—naval ports, landing grounds, etc., as well as its own. We thus obtain an intimate relationship between the activities of the three forces, the correlationship of which culminates in occupation. The army protects the naval and air bases and exerts physical pressure ; the navy secures the sea communications of the army and air force and exerts economic pressure ; and the air force provides the army and navy with information and local protection and exerts moral pressure. As moral and economic pressure take effect, the enemy's resistance is reduced, and in inverse proportion is our physical pressure increased and occupation effected. The control of these forces through their correlation is the domain of grand strategy, and the structure of the plan of expenditure and the method of maintaining them of grand tactics and of what I will call grand administration. These are the three closely related divisions of the mental sphere of war which forms the foundation of all military action.

### 10. The Study of the Mental Sphere

At the end of the last chapter I said that if we can discover the nature of the mentality of the enemy's command, then, if we work scientifically, we shall be able to discover what to expect.

The mind of man, as we know, is largely controlled by reason, and from his brain originate all his activities. In an army it is much the same. What is the governing reason of any action? We can discover this by waiting for cause and effect, and, though this method has frequently to be resorted to, it is costly, yet, when once we have ascertained the relationship between cause and effect, we shall have discovered the reason in question. By this process, by degrees we can diagnose the mentality of the enemy's command.

Another process is to examine the structure of the organ of command. What is the nature of its machinery? What can it make? Does the enemy possess an organization which can create grand strategy? If not, then we shall know that one weak link in his harness is the link which connects politician to soldier, and, consequently, by striking at the politician, either directly or through the will of the hostile nation, we may cripple the enemy's fighting forces.

What is the nature of his grand-tactical machine? Does it permit of an output of combined force? Does it link Service to Service, and weld all three Services into one force? If not, what kind of plan can it create? If we can only answer these questions we shall have gone a long way toward formulating our own plan and of discovering the enemy's weak points, his weak mental points which eventually will reveal themselves as weak physical points and weak moral points—points we should attack, and if we can foresee them, then we can plan to attack them.

We can apply this system to the study of history. For instance, we can take a campaign and link together its operations—marches, battles, etc.—and so produce a mosaic. For each operation we can by degrees deduce a reason, and, having compared these reasons, next we should turn to the brain which has conceived them and the mental machinery which elaborated them. Which is at fault? Or to which is success due? Was genius in command? Or was the organization of command defective?

Lastly, when we have made up our minds where the fault lies, we should look and see if, after the war was concluded, the enemy possessed the ability to discover it and the courage to remedy it. If not, then we can surmise that in the next war he will commit his mistakes over again—that he is, in fact, a congenital fool.

Thus, by a systematic examination of the past, can we remedy the present and prepare for the future, building up an instrument the powers of which can be expressed either by genius or normality.

# CHAPTER VII

## THE MORAL SPHERE OF WAR

Who best can suffer, best can do.—MILTON.

A man's acts are slavish, not true but specious; his very thoughts are false; he thinks too as a slave and coward, till he have got fear under his feet.—T. CARLYLE.

### 1. THE MORAL ASPECT OF WAR

CLAUSEWITZ in the third chapter of his third book writes:

> The moral forces are amongst the most important subjects in war. They form the spirit which permeates the whole being of war. These forces fasten themselves soonest and with the greatest affinity on to the will which puts in motion and guides the whole mass of powers, uniting with it, as it were, in one stream, because this is a moral force itself.[1]

It is to the great credit of Clausewitz as a military thinker that he saw the importance of the moral sphere in war. In the eighteenth century it had been grossly neglected; then came the French Revolution, which, in the form of a moral explosion, liberated the pent-up instincts in humankind, and shattered or shook every existing system of thought, including the contemporary theory of war based on Frederick's idea that the soldier is but a mechanical instrument.

Napoleon showed that he was nothing of the kind, for his system of command was not so much based on discipline as on "moral touch," or that contact between the heart of the leader and the soul of the led which makes of the soldier an animated instrument and a willing and eager partner. It was this partnership which had so long been deficient in war, and which Napoleon revealed and which Clausewitz enshrined in his book, and which many of his followers, as so frequently is the case, misinterpreted, until the moral became the only side of war.

War, to Clausewitz, "is an act of violence intended to compel our opponent to fulfil our will."[2] Physical force is the means,

[1] *On War*, vol. i., p. 177. [2] *Ibid.*, vol. i., p. 2.

and mental force is the impulse, for to Clausewitz "the compulsory submission of the enemy to our will is the ultimate *object*," the immediate object being disarmament. Of the means—namely, the physical instrument—Clausewitz writes: "The Art of War has to deal with living and with moral forces. . . . Courage and self-reliance are, therefore, principles quite essential to war."[1] This is what Napoleon realized, and this is what Jackson had in view when he wrote: "Hence the difference between a mechanic and a man of genius entrusted with the command of an army. The one operates mechanically by the impulse of fear on the slavish passions of man; the other insensibly insinuates and incorporates himself with his soldiers, forming them into heroes; . . . hence the same instruments, independent of the mechanical mode of application, move forward to victory or recoil in defeat, according to the mode in which they are animated."[2]

It is this animation which so largely constitutes the art of war, and of which it is so difficult to write. It is not one soul lighting another—this is mere fanaticism—but rather one mind illuminating many minds, by one heart causing thousands to beat in rhythm, and in a rhythm which, like a musical instrument, accompanies the mind in control. It is a union between intelligence and heart; between the will of the general and the willingness of his men; that fusion of the mental and moral spheres.

This, indeed, is a tremendous subject, and one requiring the closest study, for, though *moral* is all-important in war, it is not a thing in itself, as it is so frequently considered to be, but a link between will and action; and it is thus that I intend to view it. First, I will examine this problem from its individual side. I will attempt to extract certain moral elements of war, and explain how these are controlled and directed by a general, and then, in the latter half of this chapter, I will examine it from its collective side—the moral aspect of crowds, of armies, and the psychology of war generally. Yet by means of the written word how little can really be explained.

## 2. The Three Governing Instincts

For a moment, to return to the last chapter, so that I may establish a link. We must realize that it is our reason which enables us to discover anything. Reason to man is what force is to the universe. All universal motions are changes in force and so are all human activities directly or indirectly influenced

[1] *Ibid.*, vol. i., p. 21. [2] *A Systematic View*, etc., p. 214.

by changes in reason. Therefore, if we look upon reason as the directing force in our lives we shall at once realize that, not only must the mental sphere in which it operates strongly influence the moral and physical spheres, but that, conversely, any change in the moral and physical spheres must influence our minds and, consequently, our reason. These influences may be beneficial or detrimental, and, accordingly, so will reason be attracted towards or repelled from truth and error.

The moral sphere is the domain of the soul, ego, or " heart "—there is no just name for this element—and this, I think, alone shows how complex this sphere is. Within it lie hidden the instincts of man, and of these the strongest in war is the instinct of self-preservation, which I will examine in the second half of this chapter.

In chapter iii., when considering " The Threefold Nature of Man," I said that reason was the faculty of thinking, and that " when thoughts are fixed in one direction by a conscious impulse the result is will." Instincts, as is generally known, lead to unconscious or subconscious impulses—impulses which are not controlled by reason, and which, unless they are brought under control, may at any moment be awakened by danger, which, if not controlled, will dissipate our will-power and overthrow our reason, leaving us at the mercy of a variety of forces—fear, rage, frenzy, panic, madness, etc.

The question now arises: How are we going to fortify our will-power, how are we going to protect it so that it can withstand the shattering blows of fear? To answer this question it will help us if, for a moment, we return to the scientific method of enquiry.

Let us first observe all the instincts in man and reflect upon their nature, more especially so from the point of view of war; then let us group them, and decide how we can make use of each group.

There are many ways we can arrange these instincts, and the one I intend to adopt, and which appears to me to be a common sense one, is to group them according to the activities of man's body, namely, stability, activity, and co-operation.

Naturally I cannot here examine this question in full, as it would demand a book of its own, but I intend to examine it sufficiently for the student to grasp what I mean.

Suppose, now, there was but one man in the world, and that this man wished to continue to live in the world, what would he have to do? He would have to protect himself and he would have to assert himself. He could not live by protective means only, such as by always avoiding danger, nor could he live by assertive means only, such as would be begotten by a courage devoid of fear. In order that he may protect himself, nature has implanted in his soul the instinct of self-preservation, and, in order to assert himself, the instinct of self-assertion, and it is through the co-operation of these two that he lives; and, be it noted, the first is the base of the second, for security is obviously the first requirement of self-assertion.

Suppose that this man be given a wife, and that his desire is, not only to live, but that she should live and that their children should live. Then we find, not only a co-operation between self-preservation and self-assertion within each individual, but between each individual, which results in a give and take. In order not only that the individual may live, but that the race may survive, Nature has implanted in man's heart yet a third instinct—the instinct of self-sacrifice. A woman will protect the life of her child even to the sacrifice of her own; so in a lesser degree will man risk his life to protect his wife. These acts are not rational acts, but moral acts. As the great human trinity is man, woman, and child, so the great moral trinity is self-preservation, self-assertion, and self-sacrifice. All the instincts can be directly or indirectly classified under these three groups—the stable, active, and co-operative groups. Thus the instinct of hunger would fall under the first, of pugnacity under the second, and of love under the third. There are, of course, many other instincts; in fact, I do not think that any psychologist would definitely like to say how many there are; and, even if my threefold grouping is not absolutely correct, it possesses the value of simplicity, and, consequently, is a good hypothesis to work by.

If in the mental sphere, by a process of integration and disintegration of ideas, the scientific method enables us to arrive at the reasons for or against any suggested action, surely also in the moral sphere it will enable us to discover what is morally advantageous and disadvantageous to the control of the physical sphere of war and its elements. By the process of observation, reflection, and decision we can sort out three groups of instincts, namely, those which accentuate fear, accentuate courage, and accentuate comradeship. These three groups are essential to war. Do not let us for a moment suppose that, if we could eliminate the first group, we should fight the better for it. A man who possessed no sense of fear, no instinct of self-preservation, would fight like a frenzied maniac; that is, he would never think of protecting himself, and, consequently, would run untold and inane risks, and die the death of a fool. Again, if we could eliminate fear altogether we should have no weapon to fight with, for all physical weapons are made to instil fear. Without fear war would be a struggle of maniacs; without courage it would be a scramble of cunning cowards, of assassins who could only knife an enemy when his back is turned, and without comradeship it would be the brawl of a mob latent with panic. It is fear, courage, and comradeship which moralize war, not separately or individually, but collectively and unitedly.

Granted that these three elements are necessary to war and to scientific fighting, granted that we know their values and the value of their ingredients, then we can cultivate habits which will enable us to control, in some small way, our instincts, and which will enable us to balance and adapt them to our needs, and free our will to control our physical energy and all the activities dependent on it. Granted this freedom of will which, through comradeship, can control fear and courage, then by repetition and education we can cultivate in ourselves and our men those acquired movements which will transmute conscious associations into subconscious habits. This is, in fact, the aim of all military training.

### 3. The Relationship of Will and the Moral Elements

From this general aspect of the moral forces I will turn to the more purely military aspect, and establish a relationship between will, the final expression of the mind, and fear, courage, and *moral*, the three moral elements in war.

In peace-time, comparatively speaking, our minds are little affected by fear, but in war-time it is the reverse; consequently the direction of will-power becomes a far more difficult problem than the formulation of reasons which give will its force.

Just as a butterfly is related to a chrysalis, and the chrysalis to a caterpillar, so is will, as a physical act, related to will as a sentiment, and through sentiment back to will as a mental decision. I will now turn, therefore, from what may be called rational will and consider will as a potential rational element operating in the moral sphere, and attracted, repelled, or balanced by the elements of fear and of *moral*. Thus reason gives expression to will, will has to traverse the moral sphere before it can influence the physical, and during this journey, if reason is to rule, it should be the controller of the moral elements. To gain this control, fear must be balanced by *moral*, and, when this control is gained, not only does the soldier become a moral agent, but the will itself reverts to its rational position, and, the body being controlled by reason in its normal mental sense, it expresses the decision of the mind by a physical act of will.

Thus, if my intention to-day is to kill a certain man, and to-morrow I meet him, my will changes from a rational to a moral mood, and, once I have overcome such fear as his presence instils, the act of killing him expresses my original intention. To overcome my doubts, when he confronts me my *moral* must balance my fear, or, if I possess a low *moral,* I must rely on cunning; but of this quality I shall speak later on.

For a moment I will turn to the physical sphere, and here we are confronted by a simpler problem.

War presupposes changes in force, and particularly in physical energy. If two men wish to fight, they must expend muscular power in order to move, hit, and guard. In the first—the expenditure of force in approaching each other—*moral* must balance fear in order to allow the will to "enclutch" (to use a mechanical term) with muscle. To hit demands that *moral,* for the time being, must "demagnetize" the will from fear, and directly the blow has failed, and the hitter is placed at a disadvantage, fear must remagnetize the will so that it is able to direct muscle-power to expend itself protectively—namely, in guarding; that is, in warding off or avoiding a blow. Thus, by balancing fear and *moral* according to the circumstances in which muscle-power should be expended, the will maintains its freedom of action, and endows the muscles with freedom of movement, of which there are three moods:

(i.) Movement towards or away from the objective decided on by the reason.

(ii.) Offensive movements governed by a moralized will.

(iii.) Protective movements controlled by a will rendered prudent by fear.

When one party is at a great disadvantage, especially physically, brute force of necessity must be replaced by craft; the result is that *moral,* to a large extent, manifests as cunning, and the attack becomes a moral one—that is, an attack against the nerves rather than against the body of the enemy.

Of hunting, Jackson writes: "It prepares man for war by confirming courage or by sharpening address. If the object of the chase be the destruction of the ferocious and bold animals, the hunter insensibly acquires courage, intrepidity, and above all promptness of decision in the instant of danger. If the prey be timid and shy, he acquired address and management; for his faculties are sharpened, and his thinking powers exercised, in contriving the means of accomplishing his purpose."[1] So also in war it is the physically weaker side which exercises its

[1] *Ibid.*, p. 20.

thinking powers, whilst the stronger so frequently relies on brute force to accomplish its ends. It is, in fact, the old story of David and Goliath. Both were courageous men, but the first was the victor, for the *moral* which fortified him was intellectual.

I will now turn back to the moral side of war. In peace-time we have what is called civic control, which draws its force from peaceful morality. It is an acquired force based on certain primitive instincts. In its elementary form it is a conscious association, but in order to exert its full powers it must become

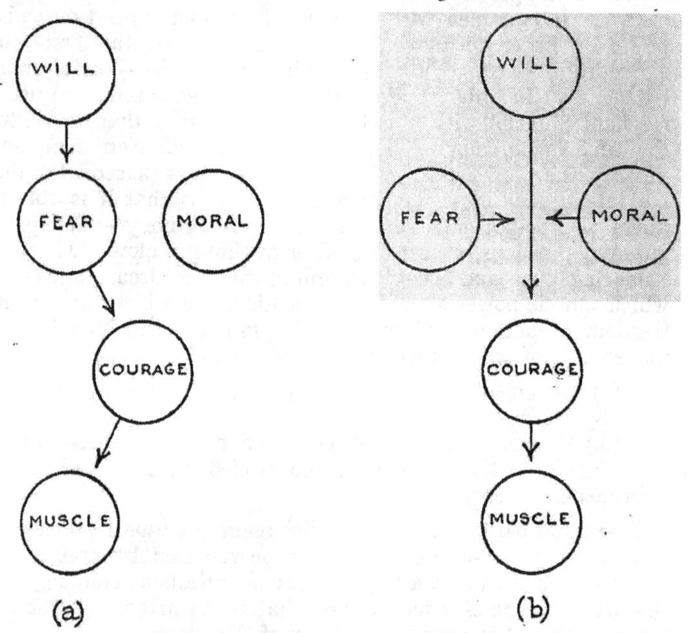


DIAGRAM I.—THE BALANCING OF MORAL AND FEAR.

subconscious and automatic. Primitive man (and still many highly civilized ones) was largely influenced by his instinct of acquisitiveness. To-day normal man does not steal, for his desire to steal has been balanced by the artificial moral reflex called honesty. In war, fear must similarly be balanced, and we balance it by means of what we call *moral*, which draws its strength from the instinct of self-sacrifice, just as fear is derived from self-preservation, and courage from self-assertion.

To recapitulate. Imagination lights up the landscape of the mind; reason takes stock of what the mind sees, and, in arriving at a decision, liberates the will which carries the message delivered to it into the moral sphere. Here it first comes under the attractive and repellent forces of fear and *moral*. If *moral* is weak, fear will block its course, as shown graphically in (*a*) of diagram 1; or, if strong, it will repel fear, and clear the way for the will to co-operate with courage, and through courage with muscle (*b*). If fear blocks the way of the will, the will will react in a direction away from danger; if, however, *moral* were to block the way, the reaction, though towards the danger, would be a very unstable one, such as expressed in rage or frenzy. It is only by balancing these two elements that we obtain a "straight" path for the will to travel along. Fear and *moral* must, in fact, repel each other sufficiently to allow of the full force of the will acting on courage, which in the moral sphere is what will is in the mental.

### 4. The Element of Fear

As I have just stated, will is balanced by fear and by *moral*, both are essential to the maintenance of will, and when they balance each other the course of will is rationally directed. We do not attempt to annihilate fear by *moral*, but to control it. If fear is under the control of the will, it becomes its most potent weapon; but, directly this control ceases, this weapon, which is a living force, not a mere inert object, turns on its wielder. To make a comparison, for fear substitute a horse. As long as the horse is under the control of its rider it is of service to him; but if it takes control he may be dashed to the ground. Control here is horsemanship; in a war it is manmastership (*moral*). Horsemanship without a horse is a useless quality, and so is *moral* without fear. By controlling fear, *moral* enables the will to execute the dictates of reason, just as horsemanship enables the will of the rider to control his horse and carry out the reason of his ride.

Fear may be moral or physical, and in a war the two are closely related. Isolation, the dread of the unknown and the unexpected, may so unhinge the soldiers' *moral* that some incident, quite unrelated to the imagined danger, may detonate his fear into panic, and, by severing his will from his reason, for a period reduce him to an irrational state. Moral fear, like a mist or fog, magnifies every danger, and by degrees it will so sap the reasoning powers of the soldier that it will create around him a phantom world which to his distorted brain is substantial and existing. Physical fear, as I think, works on opposite lines. It is not because the soldier does not see the danger that he is fearful, but because he does see it, and so clearly that he cannot avoid seeing it. If he possesses skill and weapons of equal power to his enemy he will see the danger which threatens him as his enemy sees it; if he does not, its form, though none the less true, will be exaggerated, for the degree of the danger which confronts him is directly related to his power of meeting and overcoming it.

Fear unhinges the will, and by unhinging the will it paralyses the reason; thoughts are dispersed in all directions in place of being concentrated on one definite aim. Fear, again, protects the body; it is the barometer of danger; is danger falling or rising, is it potent or weak? Fear should answer these questions, especially physical fear, and, thus knowing that danger confronts us, we can secure ourselves against it. Whilst moral fear is largely overcome by courage based on reason, physical fear is overcome by courage based on physical means.

### 5. The Element of Courage

Courage is the pivotal moral virtue in the system of war as expounded by Clausewitz. He writes: "Primarily the element in which the operations of war are carried on is danger; but which of all the moral qualities is the first in danger? *Courage*."[1] And again: "War is the province of danger, and therefore courage above all things is the first quality of a warrior."[2] And yet again: "As danger is the general element in which everything moves in war, it is also chiefly by courage, the feeling of one's own power, that the judgment is differently influenced. It is to a certain extent the crystalline lens through which all appearances pass before reaching the understanding."[3]

"Some people think that theory is always on the side of the prudent," he writes. "That is false. If theory could give advice in the matter, it would counsel the most decisive, consequently the boldest, as that is most consistent with the nature of war, but it leaves to the general to choose according to the measure of his own courage, of his spirit of enterprise, and confidence in himself. Choose then according to the measure of these inner powers; always remembering that there never was a great general who was wanting in boldness."[4]

All this is quite admirable, yet unfortunately the followers of

[1] *On War*, vol. i., p. 20.
[2] *Ibid.*, vol. i., p. 47.
[3] *Ibid.*, vol. i., p. 101.
[4] *Ibid.*, vol. iii., p. 184.

this great man misunderstood him, for replacing courage by ferocity, they established on this misunderstanding the inane theory of the *offensive à outrance.*

Jackson, who in my opinion was a profounder thinker than Clausewitz, examines this subject more scientifically. He says:

> Habits of practice give, to the soldier, such skill and management in the use of arms in the day of battle, as might be expected to be acquired by experience, in working, in unison, the separate parts of a machine of compound movement. The knowledge and ability, acquired by such experience, aided by a correct direction of powers in general movement, ensure the application of united impulse, at the proper time and in the proper circumstances of action, producing a powerful effect, and a calculable one, as depending upon a uniform rule. It is thus that experience of actual war imprints, upon the soldier, the character of veteran—a courage, arising from knowledge of things, and a consciousness of superiority in the art of applying powers. Such courage is cool and tempered: that of unexperienced troops is impetuous, blind, and headlong—liable to mistake its purpose unless plain and prominent in all its aspects.[1]

To Jackson the instinct of courage is not sufficient, any more than natural intelligence is sufficient in order to reason out the operations of war, or physical strength in order to manipulate weapons. Intelligence is the source of reason, and reasoning is a process which can be cultivated; so also with skill, and so also with courage in its military form of determination to conquer and not merely fearlessness of death. I will now examine this element of *moral.*

### 6. The Element of "Moral"

If we turn to our bodies, we find innumerable cells working on different tasks in order to maintain the structure of our organization. If we turn to society, we find individuals and groups working in the unity we call the nation. Again, in the home, though the primary instinct in man and woman is to preserve their own lives, directly children are born to them self-sacrifice replaces self-preservation. Thus whilst the individual has given us fear, the mated couple has given us something stronger than fear, namely, love, which engenders the highest form of courage, the very genius of courage; and it is on love in its many forms that the *moral* of the soldier is founded. The true soldier must love his country, and we call this affection patriotism; he must respect his leaders, and this virtue is called loyalty; he must

[1] *A Systematic View*, etc., p. 185.

have confidence in his fellows, and we call this comradeship; and, further, he must possess confidence in himself and his arms, and these are called self-respect and skill.

All these virtues, and many others, such as justice of cause, nobility of race, an honourable history, etc., must endow the soldier with a spirit which transcends all selfishness. Knowledge will help him to attain this high standard; but in the stress and turmoil of war knowledge must be backed by an intuition that, if the circumstances demand the sacrifice of his life, he must not hesitate to surrender it, so that his country may endure; just as a man or woman will risk and face death to safeguard their children. Whilst in fear is concentrated all that is brutal in man, *moral* gives to war that sublimity which raises valour to the highest of the virtues.

For the soldier to love his country his country must be worthy of his affection; to respect his officers these men must be worthy of his respect; and so we see that this virtue—*moral*—is not one which can be inculcated by the ordinary, the vulgar, methods of teaching, but one which can only be absorbed, consciously and subconsciously, by the soldier by placing him in surroundings which feed and strengthen what is of essential ethical worth within him. If the soldier feels that his officers are ever striving to preserve his life, to shield him from unnecessary fatigues, and to render his life a happy one, he will, when the occasion demands sacrifice of life, endure to the bitter end, and face the dangers and discomforts of war if only to show his gratitude—that is, his love.

Be it never forgotten that man is essentially a noble beast, for without nobility of character man would never have raised himself to be lord of the animal world. In the heart of the meanest peasant and poorest worker burns a divine spark.

Frequently we cannot see it, yet it is there. It is for us to blow this spark into a flame which will light the will of our men along the cavernous track of war, chasing the shadows from their minds, unmasking fear, mastering it, and compelling it to obedience. To obey the will of a leader is a small act, but for a man to compel fear to obey his will is a great and a wonderful act, and this compulsion is the magic of *moral.*

### 7. The Meaning of Generalship

In chapter v. I examined the structure of the control of an army, and explained how eventually this control must rest on the authority of one man, a man who possesses the power to say " Yes " or " No." There I dealt with the outer or organic restrictions and the machinery of control; now I intend to examine the moral side of this question—the ability of a general to express his power of control, when unimpeded by such artificial restrictions as councils of war, and command by conference or committee.

The moral elements, like the mental, are common to both the general and his men, but when compounded their structure is dissimilar. The general has to command, and his men, in order that he may command, have to obey. The instrument through which the general expresses his will must, therefore, be a disciplined one; that is, it must be tuned to react to reason.

In the past (and still to-day) discipline aimed at creating an instrument which reacted to the will of its leader, and the result was automatic in place of intelligent obedience. Though in certain circumstances it enabled the instrument to act with wonderful precision, when these circumstances did not exist it could not act at all, because it possessed no reason to guide it.

In the scientific training of an army the first requirement of soldiership is leadership; each man as an individual must be able to lead himself and the team to which he belongs. This leadership must be intelligent; that is, the soldier must make use of his reason, imagination, and will. He must also be able to change automatically from the active mood to a passive one, and subordinate these mental forces to the will of his leader, not as a blind force, but as a rational force—that is, a will expressing a reason or idea. This idea, the general's idea, as expressed in his plan and governed by the object of the operation to be undertaken, is his true leader, for it is not part of another man, but part of himself. The moral aim of generalship is to attain so close a contact between his reason and the soldier's reason that the two reasons fuse into one and operate as one mental force. This is accomplished by the co-operation of the will of the general and the will of his men in the moral sphere of war.

"In war men are nothing; it is the man who is all,"[1] was a saying of Napoleon's which is only partially true, and less true to-day than in his, for as the men are the implement of the general, and an animated implement, their importance needs no emphasis. Another saying, and a truer one, was: "An army is nothing without a head";[2] in fact, as much use as a bow without an archer, but with this difference—that whilst the bow is controlled by outer and physical force, an army is controlled by an inner and moral force. Jackson expresses this clearly when he writes: "A great and good general is . . . in himself an host; for his influence, insinuating itself into every member of

[1] *Correspondance*, xvii., No. 14283. [2] *Ibid.*, xix., No. 15332.

the military body, connects and binds the whole together imperceptibly, but firmly and securely. Such confidence in a leader is the charm against a panic."[1] By greatness of character a general gains command over himself, and by goodness of character he gains command over his men, and these two moods of command express the moral side of generalship.

In the turmoil of war the condition of mind of a general is the paramount factor. Has he command of himself, and through himself of circumstances, or is he lacking in this self-command? Clausewitz grasped this very clearly. He writes: "This difficulty of seeing things correctly, which is one of the greatest sources of friction in war, makes things appear quite different from what was expected. The impression of the senses is stronger than the force of the ideas resulting from methodical reflection, and this goes so far that no important undertaking was ever yet carried out without the commander having to subdue new doubts in himself at the time of commencing the execution of his work. . . . Firm reliance on self must make him proof against the seeming pressure of the moment."[2]

Here Clausewitz accentuates very clearly the value of resolution in a general, and to a general resolution is what courage is to his men. Yet the pressure of the moment may be actual and not merely seeming. Consequently resolution of itself may cause a general to act like a man galloping into a bog. Besides resolution, a general must possess a sense of caution, which is what fear is to his men, and the relationship between these two is wisdom, which is really common sense, or action adapted to circumstances.

Clausewitz, I think, leans too much on the brutal side; his general is like a charging bull, his head is well down. He possesses great strength of mind, and in place of seeing things correctly, as Clausewitz urges him to do, he refuses to see them at all; he is a magnificent animal, but not a cunning brute. If, now, to this strength of mind we can add a scientific outlook, then I think we shall obtain our ideal general.

To see correctly a general must understand the nature of the changes which take place in war. The enemy does not attack him physically, but mentally; for the enemy attacks his ideas, his reason, his plan. The physical pressure directed against his men reacts on him through compelling him to change his plan, and changes in his plan react on his men by creating a mental confusion which weakens their *moral*. Psychologically, the battle is opened by a physical blow which unbalances the

[1] *A Systematic View*, etc., p. 220.
[2] *On War*, Clausewitz, vol. i., pp. 76, 77.

commander's mind, which in its turn throws out of adjustment the *moral* of his men, and leads to their fears impeding the flow of *his* will. If the blow is a totally unexpected one, the will of the commander may cease altogether to flow, and, the balance in the moral sphere of war being utterly upset, self-preservation fusing with self-assertion results in panic.

Though the attack is one of idea opposed to idea, obviously the first step is to possess an instrument, and to deploy it so that it can withstand the physical shock; the second is to have sufficient physical force in reserve to maintain its strength; and the third is to be in a position to control the expenditure of force. Unless these things are possible, the whole stress of the battle is by degrees directed against the general until he loses control, and his army, without a head to direct it, becomes a panic-stricken mob.

This mental endurance of the general I have already dealt with in the last chapter, but it is so intimately linked with the moral side of war that I have perforce had to return to it. It is the plan which is the moral base of action, and it is the character, the greatness, and goodness in the general which sustains the plan.

To Clausewitz, besides resolution a general must possess *coup d'œil*,[1] which is attained by the "mental" eye rather than the physical. To Napoleon, a Latin, it is "to have a cool head," which never gets heated by good or bad news.[2] The quality varies according to national and racial character, but whatever it is that makes the general great, as good and worthy it must be presented to his men. "The personality of the general is indispensable," said Napoleon; "he is the head, he is the all, of an army. The Gauls were not conquered by the Roman legions, but by Cæsar. It was not before the Carthaginian soldiers that Rome was made to tremble, but before Hannibal. It was not the Macedonian phalanx which penetrated to India, but Alexander. It was not the French Army which reached the Weser and the Inn, it was Turenne. Prussia was not defended for seven years against the three most formidable European Powers by the Prussian soldiers, but by Frederick the Great."[3]

Jackson writes in a similar strain: "Of the conquerors and eminent military characters who have at different times astonished the world, Alexander the Great and Charles the Twelfth of Sweden are two of the most singular; the latter of whom was the most heroic and most extraordinary man of whom history has left any record. An army which had Alexander or Charles in its eye was different from itself in its simple nature. It imbibed

[1] *Ibid.*, vol. i., p. 50. [2] *Correspondance*, xxxii., 182-3.
[3] *Mémoires écrits à Sainte-Hélène*, Montholon, ii., 90.

a share of their spirit, became insensible of danger, and heroic in the extreme."[1]

The great general creates enthusiasm in his men by his mental and moral superiority. It is not merely success which accomplishes this, but prodigious success—success which would have been impossible without the mind of the general. Xenophon and Turenne appeal to the heart; Cæsar, Marlborough, and Frederick showed an all but supernatural skill; Gustavus, Scander Beg, and William Wallace electrify the heart of entire nations; and of Napoleon I cannot do better than quote Carlyle: "There was an eye to see in this man, a soul to dare and do. He rose naturally to be the King. All men saw that he *was* such."[2] This heroism, says Carlyle, is "the divine relation (for I may well call it such) which in all times unites a Great Man to other men." This does not explain much, but it does explain something, for it tells us that a general must possess something which is not common to his men, something which they do not possess and do not fathom. The man of normal ability is soon known to the soldier; a great general must always remain a mystery. He must never be measured; every act must appear a wonder and must rouse the emotions; it must thrill the nerves of his men and electrify their hearts. Therefore I think that originality, when coupled with a clear head and a resolute character, is perhaps the greatest gift of generalship. And to be original he must see things for himself, move amongst his men, and decide of his own accord.

In the last great war we saw no such leadership, because in place of one man controlling armies we find a staff doing so instead. It was a war run by committees and conferences, a slow-moving, inarticulate business, in which that spark of generalship which one man alone can fire, that spark which detonates the heart of the soldier and imbues him with spiritual valour, was entirely wanting. It was a truly democratic war—a Peloponnesian affair without even a Brasidas.

### 8. The Foundations of Human Nature

Now that I have dealt with the moral aspect of war, with its elements, and with generalship, I will turn to its psychological aspect, and consider in particular the psychology of the instrument. It is a complex problem involving man and men, individuals and crowds, yet in its solution is to be sought the mainspring of leadership.

[1] *A Systematic View*, etc., Robert Jackson, pp. 218, 219.
[2] *Lectures on Heroes*, Thomas Carlyle, lecture vi.

To begin with, I will ask this question: What is human nature, what is character, and what are instincts and impulses? I cannot enter deeply into this question; briefly I will answer it as follows: Character is the quality which differentiates man from man; instinct the quality which relates man to man; and impulse the product of character and instinct.

From the soldier we strive to obtain war-like impulses, and his character and his instincts are going to affect these. His instincts are common to those of his fellows, consequently character becomes a predominating moral factor in war, and one which may be cultivated, for, though certain qualities of character are inherited, others are acquired. Man is not born honest, or truthful, or loyal, yet these three virtues and many others will help to mould his character as surely as will vices. I will now turn to instinct.

In the individual, human nature is largely based on personal survival through personal striving; in the family, on family survival through propagation; and in the race, on racial survival through co-operative effort.

In the first there is a co-operation between the will and the muscles of the individual; in the second, between the desires and bodies of the opposite sexes; and, in the third, co-operative striving is directed towards united effort and common survival. The question may now be asked: Co-operation against what? And the answer is: Against death to the individual, family, or race!

Human nature is, therefore, striving against death, or, conversely, human nature is urging mankind to live. We thus obtain a threefold order—death, human nature, and life; and, as the physical aim of war is destruction, so the psychological aim is preservation, or the avoidance of destruction; consequently military psychology includes, not only the cultivation and preservation of human force, but its expenditure in war at the highest profit. Thus, the psychological purpose of war is the materialization of the human will through physical and material means in order to destroy or preserve life, the missing $x$ being death to the enemy or life to his opponent, the first being the negative, and the second the positive, values of this tremendous equation.

### 9. The Instinct of Self-Preservation

Self-preservation is the master of all life; directly a healthy child or animal is born, directly a seed begins to sprout, its one instinct or tendency is to live, and this condition remains good until death terminates its striving.

A seed in the ground will throw out its roots towards moisture, and leaves will turn towards the sun. A hare in the field will lie low on hearing an unfamiliar sound, or a bird will fly away, and man, in his own manner, will do likewise, because in all these cases it is the instinct of self-preservation which cries subconsciously within all: Avoid death; avoid the unknown; live and strive ever to this end.

From that fearful individual, natural man, I will, for a moment, turn to the soldier, for the difference is indeed startling.

What is the soldier? Right through the ages we see him leading the advance. Great nations are born in war, and decay in peace. All things strong, virile, and manly spring up during a great war; and only a few years back we saw among ourselves a whole empire gathered together to meet a common foe, each soldier possessed by one common thought—the conquest of the enemy even at the cost of his own life.

Here we have the answer to our question. It is not drill, nor uniform, badges, or weapons, which make the soldier, but that spirit of self-sacrifice for a cause which he instinctively feels he must follow, which urges him on towards a goal he may never attain, or, reaching it, may receive no further award than the knowledge that through efforts known only to himself he has added to the greatness of his country and to the security of his race. Where the civilian pays in gold, the soldier buys in blood. Where the former seeks material gain—the good things of this earth—the latter seeks an ideal which frequently can endow him with no immediate benefit. It is for this reason—the staking of his life for an ideal—that right through history, which is itself but a relation of wars, the soldier stands forth pre-eminent among the crowd of lesser men.[1]

Man being naturally fearful, whence originates this power of self-sacrifice? Again the answer is: In his nature, which is further controlled by the instinct of the preservation of the family. It is in the cradle where *moral* is born, and in the home where it is nursed into a human force. Every normal man will defend his mate, because his mate is mother of his child. She in her turn will lay down her life for his child, and so abrogate, by the highest act of self-sacrifice, her individual self-preservation for the preservation of the family. Here, then, in the family

[1] Jackson considers that it is pride of honour "which gives a character of pre-eminence to the soldier." And "Where war is undertaken in defence of liberty and national independence, it may be said to move in its highest sphere. It engenders the pride of honour; for it implies the defence of the feeble, the protection of the ashes of the dead, and the security of inheritance for those who are yet unborn" (*A Systematic View*, etc., pp. 215, 217). For a fine description of an army proud of the "honour of its arms" see Clausewitz, vol. i., p. 182.

of our primeval ancestors is to be sought the beginnings of human altruism—the affection for others, the love for little children, the sense of self-right, and race-right, and national-right, of courage throttling fear and of sacrifice scorning prudence. Here among the withered leaves and offal of man's primitive home is to be sought the foundations of society, of politics and law, and *moral* of the soldier.

Behind the soldier there stands this mystical impulse, born of the first mother, born of the first protoplasm which, dividing, lost its individuality, its desire to live, so that its species may survive.

It is this impulse which impels the soldier to do certain things so that his race may continue and prosper. Really there is nothing reasoned about this, and it cannot, therefore, be judged by rational standards—with mental pennyweights and pint pots. It is difficult to follow, as are all psychological factors, and especially those which guide and control masses of men as distinct from individuals.

The growth of the instinct of the preservation of the family leads directly to the instinct of national preservation—that impulse which, when awakened, will urge a whole nation to save its life, just as the instinct of self-preservation bids a man seek protection from danger. But, whilst the individual only seeks to save himself, the nation as a whole thinks little or nothing of the individual; and yet, thinking little or nothing, has, nevertheless, to depend for its own existence on the courage and efficiency of each human unit which goes to build it up. So we see that, notwithstanding how great and prosperous a nation may be, unless each individual, and particularly each individual soldier, is endowed with a will to win—that is, readiness to sacrifice even life for a cause—a nation must decay and perish.

### 10. The Development of Character

How can we teach the soldier to do this; how can we take an ordinary peace-loving citizen and convert him into a soldier—that is, into a man who is willing to hold back his instinct of self-preservation and sacrifice his life, perhaps for a thoughtless word of command? This is the problem we must solve if we wish to endow our men with that fighting spirit which commands success.

There are two factors we must turn to for assistance; the first is the character of man, and the second is the law of change. Character gives to us our direction; change enables us to concentrate and distribute. Certain men possess characters which are totally unsuited for war, especially for combatant work; these we must avoid, but their class is not a large one, since most men are in nature primitive, and primitive man is a fighting animal.

And now as to change. All mortal things are born, they live, and they perish; their lives are one continuous change; for no man even for an instant remains the same man. It is truly a wonderful thing to realize that we cannot raise an eyelid, breathe a breath, or utter a word, without our bodies and brains being changed. In fact, there is not a single thing which surrounds us which is not changing us, at this very moment, for better or for worse. This being so, then, because of the law of change, inseparable from life, it is possible for us to take a man, and, through his surroundings, change him from a peace-loving citizen into a soldier—that is, into a man who thinks more of an order than he does of his self-preservation.

How, by applying this law, can we best control the instinct of self-preservation? I will take an example in order to illustrate what I mean.

A child is brought up in some filthy slum, surrounded by squalor; it witnesses theft and listens to lying; drunkenness and sordidness surround it; its life and environments are one long degradation. Is it to be wondered at that this child becomes a criminal? No; for in such circumstances few children will possess sufficient force of character to win the moral battle against these influences.

In place of filth and squalor, drunkenness and theft, I will substitute cleanliness, sobriety, and honesty—the family virtues—and in place of a criminal we get a moral man. I will now add honour, patriotism, and comradeship—the national virtues—and we get the rough elements of the soldier. Suppose that these are developed by adding knowledge, skill, endurance, and pluck—the individual virtues—then we get the fighting man, the soldier, a synthesis in every sense.

We must remember this—a man's mind is being continually bombarded by impressions from outside, and, as his character changes with each shot, it is our duty to see that it changes in the right direction; for, according to his surroundings, so will man himself be, for normal man is but a walking mirror.

### II. Character, Instinct, and Impulse

Character and instinct find their expression in impulse; a sudden influence acting on the mind gives no time for reasoning, and the soldier is thrown back on his instincts and his character. If self-preservation is uncontrolled, he acts defensively, or is paralysed; but if he is imbued with self-sacrifice he will stand

and fight it out. Besides these two instincts there are three others which largely influence the soldier, namely, self-distinction, self-deception, and self-confidence.

No healthy man is willing to die or to live unrecognized, though he is willing to deceive himself in a thousand ways in order to avoid the idea of death or of obscurity. It is by stimulating his vanity that we increase his credulity at the expense of his fears and to the profit of his confidence, and thus convert a prudent, cautious being into an idealist, a soldier—that is, a man who is willing to sacrifice his life for the gaining of a cause which very frequently he does not understand. This may seem Machiavellian, but it is not so; we must take normal man as he is, and in war even stupidity is sometimes a virtue; for when we are called upon to control masses of men it is normally far easier to lead the dull than the intelligent. This does not mean that intelligence is a vice, but that masses are not suited to its useful expression. When individuals and small units are concerned, intelligence demands a fuller liberty of action, and it should be given it, for dullness here is a dangerous quality. This difference, I think, should be remembered whenever the future developments of war are considered, for on the types of armies which may be required will depend the degree of intelligence we should aim at cultivating.

It must, however, be remembered that deception and praise rapidly volatilize under the influence of acute fear, and that it is fear which, as the expression of the instinct of self-preservation controls the battlefield, and, according to the character of the soldier, urges him to do one of three things: to retire, so as to escape danger; to remain where he is, and so avoid increasing it; or to advance and clinch with his enemy, so that danger may be overcome.

### 12. Factors which Influence "Moral"

Which course he adopts depends on how far his character has been moralized—that is, on his fighting spirit, which, in its turn, depends on the conditions which surround him. These conditions must be such that, though his nerves may be assailed, his confidence in the possibility of his task is not shaken.

This confidence depends on certain factors:

(i.) Limitations to the task set.
(ii.) Ability to carry it out.
(iii.) Encouragement while so doing.
(iv.) Protection during the accomplishment.
(v.) Immunity from danger once the task is completed.

Danger, so far as it affects each individual, must be reduced to a minimum. As this is always difficult, the greater the danger the less must a man doubt his ability to overcome it. Though in war it matters much what an individual can do, it matters far more what he *thinks* he can do; consequently the art of command does not only consist in the power of enforcing obedience, but in stimulating the imagination. Frequently it happens that the soldier who believes that all is right when all is wrong is morally stronger than he who believes that all is wrong, even if his beliefs be justified.

This power of belief does not only depend on the soldier's training, or on the perfection of the organization to which he belongs, but on the loss of the sense of danger. Morally, this is accomplished by reducing his feeling of isolation and increasing his sense of security; physically, by reducing resistance through increasing the power of his weapons.

A saying we frequently hear repeated is that *moral* is to the physical as three to one, and in our turn we often repeat it quite meaninglessly. In some minds this saying of Napoleon's conveys the idea of a feud between the moral and physical means of waging war, so that two schools of thought arise—the *moral* and the *matériel* schools. The first asserts that *moral* is more important than weapons, and the second that perfection of *matériel* is the most potent factor in war.

In my opinion, both schools of thought are wrong, because they base their ideas on a division between the moral and physical spheres of war. No such division exists, any more than it does in man himself. The heart is not superior to the body, or the body to the heart. Together these two form an integration which cannot be separated, and, as the body gives expression to the will, and, through the muscles, protects the brain, so do the physical means of war give expression to the moral, and protect *moral* itself. Consequently if Napoleon's dictum be true, and the *moral* is three times as potent as the physical, then logically we should not leave a stone unturned to obtain all possible superiority of physical means so that our *moral* is given the very fullest security. In the past, so I hold, we have thought far too much on the lines of guts *versus* guns, and when I come to discuss the physical sphere of war I will show that this conception is a fallacious one, and that there is no *versus* in the question. I will now return to the subject of this chapter.

An unlimited objective requires unlimited endurance; this is impossible; consequently the task to be accomplished must be within the mental and physical limitations of man. These powers do not only depend on preparation and training before battle, but on support and protection during it. Thus men will continue to advance if they know that they are being followed. Their self-deception urges them to believe that the moving masses behind them are immediately protecting them.

This, of course, is not so, for their protection is probably being provided for by invisible guns in rear. The support here is purely moral; it stimulates the nerves of the attackers by reducing their feeling of isolation, just as the bursting shells in front of them, by reducing the enemy's resistance, are physically enabling them to move forward.

The instinct of self-distinction urges men on, for public applause is the greatest of all trinkets, and it would be a shameful thing to lag behind whilst countless eyes are following the advance. Further, it would be a dangerous action, for behind them stands the inexorable law of the soldier which requires certain death for uncertain courage.

Ultimately the instinct of self-preservation, which has filled their hearts with an almost uncontrollable fear of individual danger, explodes into the frenzy of revenge, once the distance between them and danger is so reduced that to fall back would be to commit suicide. Collectively men "see red"; their reason vanishes, their self-deception disappears, self-distinction is forgotten, their whole being crystallizes in one word—kill—or truer, perhaps, in one word—murder, for the bayonet knows no pity.

If complexities arise in the physical struggle of battle, how much more so is this the case when we enter the psychological struggle of will against will, of nerve against nerve, of impulse, of sentiment, and of instinct. Round this struggle, between the souls of men, gyrate success and failure; for, whatever his weapons, his means of movement, and methods of protection may be, ultimately we come back to man—the frail, fearful, yet cunning creature whose supreme aim is life, whether in the peaceful field of trade or among the death-groans of the battlefield.

### 13. The Character of the Crowd

From the individual I will now turn to a mass of individuals, for the understanding of crowd psychology is the foundation of leadership, which in war is not only complicated by the instability of the crowd "mind" as affected by danger, but by the continuous change of the component parts of the crowd itself due to sickness and casualties in the field.

There are two types of crowds—the heterogeneous and the homogeneous—each of which, under a strong impulse, may become psychological; that is to say, it may act like an individual. Thus two men of different education meeting in the street form the smallest type of heterogeneous crowd, two soldiers or doctors, etc., the smallest type of homogeneous.

In both cases there is a relativity of thought, but, whilst in the first there is nothing in common in the crowd except the instincts of each individual to bind this relativity into a unity, in the second case a denominator exists. Ultimately we find that a nation forms a great mass of homogeneous crowds floating in a heterogeneous human vehicle, the whole controlled by a national "soul," the strength of which depends on the mental homogeneity of the mass itself.

In appreciating the crowd, first we must realize that the crowd "mind" is not the average of the minds of the individuals which compose it, consequently intellect counts for next to nothing in a crowd; secondly, that the common element in each mind—self-preservation, and all that self-preservation includes—counts for much. Thus, taking twenty men, the individual qualities may be $2a$, $4b$, $3c$, $1d$, $3e$, $2f$, and $5g$, but the common quality—fear—will be $20x$, consequently the human spirit will overcome individual character and ability. We find, therefore, that the combination of many minds results in the creation of a crowd "soul" which, though related to each individual soul, is uncontrolled by any rational thinking organ, for the "mind" of the crowd itself is completely dominated by it.

When we analyse the crowd we find that it is swayed by the voices of the past, and that, accepting it as an entity, we discover that that part of it which I have called its "mind" is swayed by that part of it which I have called its "soul," and that this "soul" is dominated by the instincts.

In certain circumstances the conscious personality of the individual evaporates and the sentiments of each man are focused in the same direction. A collective "soul" is then formed, and the crowd becomes a psychological one, and henceforth acts like an irrational individual in place of like a mass of separate rational individuals. The character of the crowd is now determined by certain well-known conditions:

(i.) Its feeling of being invincible, resulting from numbers.

(ii.) Its liability to be persuaded by suggestion, due to its inability to reason.

(iii.) Its instability, due to its liability to mental contagion through suggestibility.

As conscious personality evaporates, subconscious personality forces itself uppermost, so that, directly an idea is suggested, by contagion all agree to it, and, through the sense of invincibility, all set to work to carry it out. The crowd becomes, therefore, a mere automaton under the will of the suggester, and, through lack of intellect, its acts are always unbalanced and extreme—lower or more exalted than the individual's, according to the nature of the suggestion it has received. The crowd is always latently mad, and its study is virtually one belonging to mental pathology.

The special characteristics of a crowd are its impulsiveness, changefulness, and irritability. It is slave to its impulses, and cannot control its reflex actions. It cannot understand restraint, for it lacks understanding, and the greater its size the more pronounced becomes this loss of power. Its normal state is fury; it is credulous; it is incapable of observation, and it is easily hallucinated; it blindly follows example, and it falls an eager victim to such as use exaggeration, affirmation, and repetition as their tools.

Ruled by its sentiments, all ideas are either accepted or rejected *en bloc*; the crowd therefore lays down the law, and is utterly intolerant. Under weak authority it revolts; under strong it acts with the most debased slavishness; it may be noticed, therefore, that, according to the character of their rulers, crowds pass alternately from anarchy to servility and back again.

The factors which govern crowds may be divided into three classes:

(i.) Distant factors: race, religion, traditions, education, and customs.

(ii.) Immediate factors: images, catchwords, formulæ, and irrational statements.

(iii.) Future factors: promises—in one word, Eldorados. On words masses of men rapidly become intoxicated.

To carry a crowd forward to some desperate deed, all great demagogues have worked on its "mentality" by means of suggestion, the strongest form of which is personal example based on prestige—that is, on accumulated renown—for without prestige affirmation, repetition, and exaggeration lack that electric attractiveness which concentrates the sentiments and emotions of the crowd.

### 14. The Co-operative Group

A heterogeneous crowd, as I have explained, is a mass of individuals governed by uncontrolled desires which obliterate the individual will; the will is, in fact, surrendered to impulse.

In a homogeneous crowd the mental disintegration of the individual will is slower, unless it be given a definite direction, when the will is endowed with a psychological impulse.

In homogeneous crowds, such as armies, the will of the individual is not so much surrendered to impulse as subordinated to command; it is not effaced, but directed. The mental organization of a co-operative group differs from both of these crowd-forms, for in place of either surrender or subordination of the wills of the individuals these wills are brought into the closest co-operation, and contribute to the growth of purposeful thought.

In the heterogeneous crowd there is a persistent jarring between agreements and differences; in the homogeneous there is a concentration on agreement; but in the group there is a harmonization of the differences, so to speak—the opposites mate and give birth to creative thoughts. It is by overcoming differences that the group learns to live together as a united whole in a state of co-operation.

In an army this unifying group-spirit should control all its parts as groups, and ultimately as one group. That is to say, a section of ten men should not only be endowed with a sectional group-spirit, but this sectional group-spirit should form part of a platoon group-spirit, which, in its turn, forms part of a company group-spirit, and so on through battalion, brigade, division, corps, and army, until it forms part of the national group-spirit itself—the ultimate group. Only by such a process of integration can unity of will, and, consequently, of effort, be attained. In such a group, to attack one individual is to attack the whole group, which moves as one man—an articulated whole in place of an undifferentiated mass.

The strength of a group does not lie in its numbers but in its psychic force, which draws its power, not from the instinctive similarities in the individuals composing it, but from the voluntary harmonization of their differences.

This psychic force attains its highest freedom of action when a complete relationship has been established between the individual wills. This relationship is dynamic; it cannot possibly be static, since the law of change produces a new crop of differences immediately an old one has been reaped. The process of the interpenetration of the individual wills into the group will is, therefore, continuous; it can never cease; and it is this continuity of progress which gives its impulse to creative thought. The universe of mind is never conquered, for directly one world is subdued another rises bright on the horizon, which, in its turn, must be explored and won.

The simpler the organization of the group—that is, the fewer its differences—the greater becomes the liberty of thought and action of each individual composing it. In the crowd these differences are being perpetually cannoned off one individual against another, and consequently give rise to much friction. A condition which is affected by friction is one lacking in freedom, for it is hedged round by numerous obstacles.

In the crowd, men develop through an incessant struggle in which the fittest survive; in the group, survival is not attained so much by competition as by co-operation—that is, through the art of learning how to live and work together. It nevertheless must not be forgotten that, however perfect may be the organization of a group of men, in essence it is an artificial organization, its only natural prototype being the family. Its foundations are shallow, and it will probably take many generations of groups before they sink deeper, and many hundreds, possibly thousands, before the group-spirit will have grown sufficiently strong to rule the primitive human instincts which control the crowd. This is a most important fact to bear in mind when considering the stability of the military group, an organization which has never as yet been scientifically formulated. Soldiers have hitherto been organized in homogeneous crowds, and as such I will now examine them.

## 15. The Military Crowd

Turning to the military crowd—that is, any unit of drilled men—we find that it is what Gustave le Bon terms a psychological crowd—that is, a mass of men dominated by a spirit which is the product of the thoughts of each individual concentrated on one idea. If this idea be the "will to win," then the result is that the spirit of the crowd becomes an all-impelling force, urging it on as long as the individual thoughts are concentrated or focused by this will. Should, however, these thoughts be disorganized by a sudden calamity or surprisal, then the natural instincts will intervene, and the will to win will be replaced by the instinct of self-preservation. However perfectly trained a body of soldiers may be, it always tends to become once again the crowd. The power which prevents it doing so is its *moral*. So we find that, as the heterogeneous crowd is swayed by the voice of instinct, a well-ordered army—that is, a homogeneous and psychological crowd—is swayed by the voice of training, uniformity of environment having created within it a uniformity of character and spirit.

In a crowd each man surrenders his personality to his leader.

In an army each soldier subordinates his will. Herein is to be found the quality which differentiates the soldier from the civilian who, as one of a crowd, has little or no power at all, and who obeys on impulse and not on purpose.

An army we find, therefore, is still a crowd, though a highly organized one; it is governed by the same laws which govern crowds, and, under the stress of war, it ever tends to revert to its crowd-form. Our object during peace-time consequently is to train and organize it in such a manner that during war this reversion will become extremely slow; in other words, we should aim at adding to each individual the quality known as *moral*, so that, when intellect and reason fail, man is not ruled by his instincts and sentiments alone, but by his *moral*, which has become part of his very nature.

Suppose that these moral forces are represented by $y$, we then find that as the individual qualities, the $a$'s, $b$'s, and $c$'s, evaporate, the common quality, $x$, though it may push itself to the front, is, nevertheless, kept within bounds, directed and controlled by $y$—the common *moral* of each individual as well as of the crowd in its entirety.

## 16. The Psychology of Battle

I have now dipped somewhat deeply into the psychology of war, and all that remains for me to do is to weave what I have said into that complex psychological crisis of war which is called battle. The process of doing this is complicated by the fact that man must be considered, not only as an individual, but as a being affected by the psychology of a mass of individuals. In himself man is a separate cell in the military body, but, like a cell, he cannot live apart from this body, for he is affected by all the other cells, and on their moral health depends his own.

In this psychological struggle we start with known conditions: the mentality of the commanders, leaders, and men of an army. We realize from the outset that these conditions are most unstable, even amongst highly trained troops, and that this instability will begin to manifest itself through the sense of approaching danger, even before the first shot is fired. Then this danger, from a mere phantom, materializes into the tyrant of the battle-field as the first shot whistles overhead. There is the will to win, the *moral* to endure, and the sapping of the moral forces through fear. Woe to that army which has not cultivated the first two in days of peace; woe to the commander who has not only endowed his men with the spirit of the justice of their cause, but has failed to arm them with the most potent weapons, means of protection and of movement, so that confidence in victory, through superiority of equipment has become an instinct in the souls of all.

If the " mind " and " soul " of an army be strong in its strength, then its endurance will be high ; but if, in spite of all its gallantry, men be mown down by thousands, then every shot which shrieks overhead, though it may do no harm physically, inflicts a moral wound. A man is killed ; his fellows seek protection ; some surge forward, others remain behind. *Moral*, the most volatile of spirits, is evaporating under the blast of fear, that grim tyrant who ultimately whispers in the hearts of all : " Thus far, but no farther ! "

As the battle bursts into flame, creative reason holds control or is lost ; imagination rattles the dice of chance and the man obeys, or, like an animal hunting another, acts on his own intuition. Self-sacrifice urges men on ; self-preservation urges men back ; reason decides ; or, if no decision be possible, sense of duty carries the will to win one step nearer to its goal. So the contest is waged, not necessarily by masses of surging men, but rather by vacant spaces riddled by death.

According to the preponderance of *moral* or fear is homogeneity of mind and determination of will maintained or lost. Little crowds fill the battlefield, each with its own little soul trembling before its immediate future. Some advance lethargically, some with enthusiasm ; some watch others, and act in accordance with each other's impulses. The spheres of action are now revolving ; are the leaders still individuals, or have they lost their identity in the crowd ? If so, will some heroic soul re-establish it ? For in the leader lives the impulse to move.

A wounded man shouts, " Are we downhearted ? " and the little crowd surges forward, led by the phantom engendered by his cry. Then gallantly a man sacrifices himself, and again the crowd moves on, impelled by example, by rage, and by revenge. Thus is victory suggested and the will to win revived.

Then some act, frequently unknown to the crowd, tells that the victory is won. Group after group of fighters take up the unheard call, and the man who but a moment before was one of many—an individual without identity—suddenly materializes into human form. Such is the psychology of battle—a climax and an anti-climax, and yet a climax once again. Fear magnifying and rage blinding. A struggle between the bestial and human, between self and self-sacrifice, and then the ultimate relief that danger has been vanquished, that the fields are green, and that life is sweet to live.

### 17. The Study of the Moral Sphere

We talk a great deal about *moral* and the will to win, yet of all virtues they are the least susceptible to talk and the most to action. Moral force is not like electrical energy; it cannot be stored up in batteries and sold by the kilowatt or any other commercial measurement. Man himself is the battery, and his willingness and instincts are the poles. We have got to link these up by action, both mental and physical, so that, when the soldier is called upon to act, he may act rationally, courageously, and skilfully. Normally we mistake stubbornness or cheerfulness for *moral*; we might as well suppose that oxygen and hydrogen are water; they are not, though they may become water; so if we act correctly may we also become moral instruments.

To ascertain the moral value of an army is of the highest importance in war; why, then, not ascertain it in peace-time, so that we may learn, now and to-day, what to expect of it when war breaks out? Frequently we are told that war is a matter of two wills in opposition; then the supreme question is, What is the respective value of each of these two wills? Though it is difficult to answer this question, it is not impossible to set about seeking an answer. The body of man is strongly influenced by his physical surroundings, so also is his soul influenced by his moral surroundings. What are they?

What is the discipline of an army, and especially the discipline of its officers? Is it based on blind obedience, or does it aim at expanding the intelligence and of stimulating self-command? Is liberty of thought and speech allowed? Are officers permitted to express their opinions; are they educated to respect merit, or merely to acquiesce with senility? Are officers promoted because they are able, or because they are old? Are they rewarded for possessing critical constructive minds, or are they merely pushed on like pegs on a cribbage board? All these and many other questions will tell us the moral worth of an army.

Does fear predominate, or does courage? Is will free to act? Is *moral* the magnetism between will and heart, the idea in the head of one man and the willingness in the soul of another, or is it a mere copy-book precept—a shibboleth? To answer these questions we must watch the officer and the man, and above all the working of the system, and, if we think that it is defective, we must criticize it openly, so that it may blush at our criticism, for criticism is our mental hoe.

Every manual tells us that we are preparing for a war of the first magnitude, but against whom? Nobody can tell; but this should not dishearten us, for we know that the number of our formidable adversaries is limited, and we also know that the moral mainspring of each army is the character of the nation to which it belongs. If we take the trouble to understand what these characteristics are, then we shall be able to judge the tension of these mainsprings, and, once we know what the respective tensions are, we shall be able to chart out a moral map for each nation, which will give us moral direction in war. Given such a map, we shall not only be preparing for a war of the first magnitude against some unknown adversary, but against each knowable one, irrespective of whom it may be. This is how we should study the moral sphere of war. To keep on repeating like a mantra yogi, that the moral to the physical is three to one, and to do nothing, is about as helpful as saying that the moon is made of green cheese. Does the system we are examining, whether our own or that of another nation, give preference to ability? Does it attempt to foster intelligence and to discover moral knowledge? If it does, then is it a good system; if it does not, then it is a criminal one, for normally it is preparing the army in question, not to win, but to lose the next war.

# CHAPTER VIII

## THE PHYSICAL SPHERE OF WAR

The first ground handful of nitre, sulphur, and charcoal drove monk Schwartz's pestle through the ceiling: what will the last do?—T. CARLYLE.

### 1. THE PHYSICAL ASPECT OF WAR

IT is in the physical sphere of war that we find the most pronounced differences to peace, for war is pre-eminently a physical struggle for mastership in which the moral conventions of civilized nations are temporarily set in abeyance. So powerful is this final manifestation of force that even to-day it still obscures the purpose of war, and, in the mind of the average soldier, replaces the political object by one of a purely military value.

Destruction of the enemy's physical strength is the canon of the physical school of war; to the moral school, it is the destruction of the enemy's will. I have touched upon the views held by these two schools in my last chapter, and for a moment I will return to them, for, unless a true relationship is established between the moral and physical spheres, the soldier is apt to go astray, as so many soldiers in the past have done.

As a base of argument I will quote a passage from Marshal Foch's *Principles of War*. He writes:

> "Ninety thousand vanquished men withdraw before ninety thousand victors merely because they have had enough of it, and they have had enough of it because they no longer believe in victory, because they are *demoralized*, because their *moral* resistance is exhausted" (General Cardot) (merely *moral*: for the physical situation is the same on both sides). It was with this in his mind that Joseph de Maistre wrote: "A battle lost is a battle one thinks one has lost; for," he added, "a battle cannot be lost physically." Therefore it can only be lost morally. But, then, it is also morally that a battle is won, and we may extend the aphorism by saying: A battle won is a battle in which one will not confess oneself beaten.[1]

This is magnificent, but it has little to do with the reality of war; in fact, it is common nonsense.

[1] *The Principles of War*, p. 286.

To say that "a battle cannot be lost physically" is to ignore the greater part of the history of war. Take the following two cases and examine them.

(i.) I meet a man to whom I intend to give a sound thrashing. I refuse to be beaten, but, nevertheless, he knocks me down and beats me, because, as it happens, he is twice as strong as I am, is more lucky, or more skilful.

(ii.) Next time I meet him I intend to kill him. He rushes at me, but not for a moment do I lose my confidence in victory, because I pull out of my pocket a pistol and shoot him dead.

In the first case muscle wins in spite of will; in the second, the will of my adversary has no possible chance of winning. Therefore to say: "A battle won is a battle in which one will not confess oneself beaten" is absurd, for all it can mean is that if both sides are in all respects equal, save in will power, then the most determined will win, or if unequal, and the numerically stronger side is composed of cowardly soldiers, then the smaller and more courageous side may win. This absurd doctrine—military witchcraft of the lowest order—very nearly led to the extermination of the French armies in 1914.

That such a doctrine could ever have been accepted by intelligent men is amazing, seeing that Clausewitz, the high-priest of the modern theory of war, had clearly stated:

> Courage and the spirit of an army have, in all ages, multiplied its physical powers, and will continue to do so equally in future; but we find also at certain periods in history a superiority in the organization and equipment of an army has given a great moral preponderance; we find that at other periods a great superiority in mobility had a like effect; at one time we see a new system of tactics brought to light; at another we see the art of war developing itself in an effort to make a skilful use of ground on great general principles; and by such means here and there we find one General gaining great advantages over another.[1]

This is common sense. In brief, *moral* multiplies physical force, and physical force multiplies *moral*. It is not only necessary to imbue the soldier with the highest *moral* by careful training, but also to furnish him with the most effective weapons, means of movement, and means of protection, and to teach him how to make the most skilful use of these means, so that he may safeguard his *moral*, in order that this *moral* may fortify his offensive and protective actions.

[1] *On War*, vol. ii., p. 5.

Mental force does not win a war; moral force does not win a war; physical force does not win a war; but what *does* win a war is the highest combination of these three forces acting as *one* force. Do not let us, therefore, belittle physical force, for it is an essential of this trinity, and all other forces are as nothing without it. To Shopenhauer the world may well be "will and idea"; but to the soldier war is very largely a matter of blows, and, if he does not believe in them, then he will get his head cracked, and, if he only believes in them, then he will die of a moral arterial sclerosis. Carlyle cries: "Feeblest of bipeds! Three quintals are a crushing load for him; the Steer of the meadow tosses him aloft, like a waste rag. Nevertheless he can use Tools; can devise Tools: with these the granite mountain melts into light dust before him; he kneads glowing iron, as if it were soft paste; seas are his smooth highway, winds and fire his unwearying steeds. Nowhere do you find him without Tools; without Tools he is nothing, with Tools he is all."[1]

These are words of wisdom, and, in the next war, one of the supreme questions will be: who has the best tools? For it is the better weapon which more efficiently expresses will and *moral*, and more effectively protects them.

## 2. The Physical Elements of War

I have laid it down that the elements of force are stability, activity, and co-operation, and I have shown that in the mental sphere these elements are represented by reason, imagination, and will, and in the moral sphere, by fear, *moral*, and courage; now I will turn to the physical sphere.

In the normal pursuits of peace, as I explained in chapter iii., man's desire is to protect himself, and he does so through his power to work and ability to move. I have also pointed out that there is no intrinsic difference between peace and war, the difference being one of degree. Obviously, if fear is an essential element in war, man must protect himself; and, if courage is another, he must be imbued with an offensive spirit, and, obviously, he must be able to move. We thus obtain three physical elements of war—namely, protection, offensive action, and movement. The first is the stable base, the second the active, and the third the co-operative element.

In chapter iii. I showed that physical energy was expressed by the muscles of the body, and that these could either construct or destroy. In chapter v., when discussing the instrument of war, I have shown that Marshal Foch considers that all systems of

[1] *Sartor Resartus*, Thomas Carlyle, chap. v.

tactics should be based on "resisting power" and "striking power"; this is an idea which may be considered as universal in war. For instance, in Balzac's *Contes Drolatique* we read of a certain Captain Cochegrue of whom it is related: "Dans les grosses batailles, il taschoyt de donner des horions sans en recevoir, ce qui est et sera toujours le seul problesme à resouldre en guerre." ("In great battles, he endeavoured to give blows without receiving them, which is and always will be the only problem to solve in war.") And why? Because the resultant is liberty of movement, and, as Frederick the Great said, "to advance is to conquer!"

What has not been so universally accepted is the relationship between the natural and artificial means of fighting. I have shown that the three elements of force find expression in the structure of the military instrument in the form of protective troops, combat troops, and pursuit troops, and consequently, since these three types of troops no longer use their fists and teeth and solely their feet for protective, offensive, and mobile purposes, but, instead, weapons, means of protection, and means of movement, the first two of which are in nature mechanical and therefore artificial, and the last are rapidly becoming so, these artificial means should bear a distinct relationship to the elements they are intended to express. Lloyd is the only writer I know of who definitely grasped this relationship; he says: "Weapons should express force, agility, and mobility." And in his opinion an army is not complete unless it includes infantry, cavalry, and light infantry.

In most armies we see weapons evolving on no rational plan. New arms are invented and introduced without a definite tactical reason, and without a definite relationship to structure, maintenance, and control. Old weapons are maintained; the old and new are mixed irrespective of their elemental values. Proportions are not logically arrived at, but are the outcome of ignorant opposition on the one side and enthusiastic aggressiveness on the other. The whole process is alchemical, is slow and costly and inefficient; ultimately trial and error wins through. Thus for a hundred years we find the French knights charging English archers; for another hundred years or so, cavalry charging musketeers and riflemen; and I suppose we shall see for yet another hundred years infantry charging tanks. What for, indeed what for? Not to win a battle, for the impossibility of this is obvious to a rhinoceros. No; but to maintain the luxury of mental indolence in the head of some military alchemist. Thinking to some people is like washing to others. A tramp cannot tolerate a hot bath, and the average general cannot tolerate any change in preconceived ideas; prejudice sticks to his brain like tar to a blanket.

The three physical elements of war are moving, guarding, and hitting. In the unarmed fighter this is actually so; but in organized armies soldiers make use of material means to accentuate and economize their power of movement in all its moods. In order to hit they use weapons; to guard they use various means, such as cover by ground and armour; and to move they also use various means—horses, elephants, lorries, tanks, aircraft, etc. Normally, when speaking of the physical elements of war, I shall call them movement, protection, and weapons, in place of power to move, to guard, and to hit, or mobility, protective power, and offensive power.

### 3. The Element of Movement

Like the mental and the moral, the three physical elements are so closely related that to separate them is practically impossible, for the utility of weapons and protection depends on movement, and, in war, movement must have some offensive purpose, or one indirectly connected with fighting, and this movement must be protected if force is to be economized.

All physical movement depends on muscle-power. A man may ride a horse or be conveyed in a chariot or a tank, yet these means do not cancel the expenditure of physical energy, for they only economize it.

There are three forms of movement—human, animal, and mechanical; there are three vehicles of movement—earth, water, and air; and there are three dimensions of movement—one-dimensional, such as movements along roads and railways; two-dimensional, such as movements over land and water areas; and three-dimensional, such as movements under water and through the air. Since the advent of the tank, submarine, and aeroplane, the two last-mentioned dimensions are assuming an importance which will undoubtedly revolutionize warfare.

There are also three types of military movement—strategical, tactical, and administrative. Tactical movements, which are the ultimate aim of strategy and administration, may be divided into protective and offensive movements. The first I will call approach movements and the second attack movements. During the former the one thought of the soldier is to prevent himself from being hit, and during the latter it is to hit his enemy. The more he can hit the less he will be hit; consequently, indirectly though it may be, not only is the whole action protective in character, but it becomes more and more secure as the offensive succeeds; the approach persistently economizing the forces of the attack so that the attackers may, as far as it is possible, retain their initial strength, or increase it.

From this it will be seen that any idea of thinking of the offensive and the defensive phases of war, battle or fight, as separate and distinct acts is absurd, for these two acts form the halves of the diameter of the tactical circle, the circumference of which is the fight. They are, in fact, the positive and negative poles of the tactical magnet called battle.

When I deal with the principles of war I shall have occasion to enter more deeply into this subject; meanwhile, if we always remember that the object of all attack movements is to develop weapon-power against an enemy, and of all approach movements to prevent the enemy developing weapon-power against us, we shall at once realize that, when we are not attacking—and by attacking I mean using weapons offensively—we are approaching, even if we are sitting in a camp 500 miles from the battle-front. If we remember this—and for the soldier it is one of the most important things that he should remember—we shall never be surprised, and surprise to-day is far easier to effect than in the past, since aircraft can almost as safely attack back areas as front lines. The true appreciation of the approach and the attack carries with it the maximum of security and offensive power. These can never without danger be divorced.

Rising from battle tactics to campaign tactics, the same idea holds good. We are confronted first of all by the strategical movements, and secondly by the tactical. In brief, the whole of strategy consists in placing an army, or the various parts of an army, in such positions that tactical movements may be carried out with the greatest economy of force. Whatever we do, we must economize the expenditure of force. This is a point I shall frequently repeat, as it cannot be repeated too often.

### 4. The Element of Weapons

Offensive intent is expressed by means of weapons, and in organized and civilized warfare man cannot economically protect himself without them. Weapons have three purposes: to kill, to injure, and to terrorize. There are three kinds of weapons: weapons for thrusting, for hurling, and for asphyxiating. The first I will call shock-weapons—such as the lance, sword, and bayonet; the second missile-weapons—such as the arrow, bullet, and shell; and the third chemical weapons—such as gas and toxic smokes. Other weapons can be added to these, such as the club for stunning and germs for spreading disease; but, generally speaking, we need only think in terms of two types, according to the means used to move them; namely, those wielded by man and those discharged by mechanical or chemical force.[1]

In primitive warfare hitting and hurling weapons were combined in a chipped stone, which could be used as a shock-weapon when held in the hand and as a missile-weapon when thrown. To throw a stone is a protective act, which, if the projectile hits the man it is aimed at, may prevent him approaching to shock-distance. At shock-distance brute force predominates, and skill is reduced to a minimum; consequently the whole process of organized warfare has proceeded along the straight line of obviating the rough and tumble of body-to-body fighting—the dog-fight of battle. So much has this been the case that to-day we find, because of the invention of automatic weapons, the physical assault, as it was conceived a few years ago, is almost dead; and it can scarcely be doubted that, when the day arrives in which the bulk of our automatic weapons are protected by armour, the bayonet charge will be as impracticable as one Dreadnought ramming another.

Here I will not, however, pursue this future possibility, for existing weapons provide ample means of illustrating my argument.

As the object of battle is to destroy the enemy's strength, which is generally accomplished by clinching with him, or by threatening to clinch, the infantryman's offensive weapon is the bayonet, and as long as circumstances permit him using the bayonet this fact remains true.

His bullet is his protective weapon, because of its ability to secure the advance of the bayonet. Thus it will be seen that whenever two weapons of unequal range of action are employed, the one of longer range is always the protective weapon, and the one of shorter range the offensive weapon, and, even if three or more weapons are used, this holds equally good for all. Thus though field-guns, when covering a rifle-attack, are acting protectively to the rifles, they are acting offensively to the heavier guns in rear of them, though these heavy guns are simultaneously acting protectively both to them and to the rifles.

It may be considered that this is a purely academical problem, yet it is not so. Its full appreciation, in fact, forms the

[1] Of weapons Clausewitz writes: "Of all weapons which have yet been invented by human ingenuity, those which bring the combatants into closest contact, those which are nearest to the pugilistic encounter, are the most natural, and correspond with most instinct." Consequently from this he deduces the fact that the less the hand-to-hand fight takes place in war the less brutal warfare will become, for it is instinct which renders it brutal, and not weapons (*On War*, vol. iii., p. 250).

backbone of the attack, from which the whole battle organization, like ribs, radiates. From this appreciation may be deduced a tactical rule of high importance, namely:

In all circumstances missile-weapons must be employed to facilitate or ward off the shock.

And even if shock-weapons entirely disappear from the armoury of war, in spirit this rule will hold good in the following form:

In all circumstances the longer-range weapons must be employed to facilitate or ward off the employment of the shorter-range weapon.

The soldier must not only never forget this rule, but it must so completely dominate his thoughts that its application becomes instinctive, for it forms the foundations of fire-supremacy, that crucial act of the attack, the paralysing of an opponent's power to hurl, so that he may be hit, and his strength destroyed.

Every missile which can economically, that is effectively, be thrown, must be thrown. The soldier must not only think, but live and act in terms of fire-supremacy: for it is his sword and his shield, upon which his tactical life depends.

I have called the above tactical act a rule because, in my opinion, it is open of exceptions. Soldiers may on occasion be equipped with an offensive weapon of so small a value that for practical purposes it ceases to be a weapon at all, or else in battle they may be faced by an opponent so indifferently organized and trained that they can destroy him at long range without the necessity of clinching with him. Thus, at the second battle of Ypres, our rifles and machine-guns were rendered temporarily impotent by the use of a comparatively short-range weapon—gas; and at Omdurman the bayonet was of very little value, since the Soudanese could with ease be destroyed by rifle-fire.

Having now shown what an important part protection plays in movement and the use of weapons, I will consider it in itself.

### 5. The Element of Protection

The first fact which strikes us in life is that the instinct of self-preservation demands protection in one form or another, and the second, that protection demands activity, or resistance, or, better still, the two combined.

If we examine Nature, we at once see that so far as things living are concerned, nine-tenths of their activities are in character protective. In the animal world, the summit of which reaches to man, we find every type of protection being sought after and applied.

The tiger seeks security through offensive power; the lobster through its armoured shell; the cuttle-fish through emitting a "cloud of ink"; the skunk through a nauseating stench; the chameleon through a change of colour; the stick-caterpillar through its ability to represent a twig. The ostrich is supposed to hide its head beneath the sand, and it is alleged that sometimes man raises his above mere imitation, and, gazing into the future, sees the form of events that are to be.

Few studies are more profitable to the soldier than that of natural history, which is an unbroken relation of wars. This fascinating study I cannot pursue here, so I will turn to the element of protection.

The defensive has very little to do with holding a position, for it is just as much part and parcel of every forward movement as of every retrograde one. Static warfare is offensive warfare localized, the aim of both sides being quite as much to win as to avoid being defeated. A purely defensive (secure) war means that the object is to return to the *status quo* before the war began; consequently that the war has lost its meaning, for to wage war and return to the *status quo* is but to squander human energy.

I have already pointed out that the bullet protects the bayonet, and that the approach secures the attack, both these forms of protection are indirect; that is to say, they do not ward off blows, but, in place, impede the enemy from delivering them, either by inflicting blows or by rendering the target invisible or difficult to hit.

Besides the numerous indirect means employed to protect the soldier, a number of direct ones have been used, such as armour, earthworks, fortifications, and gas-respirators. Again, all these means of economizing hitting-power may be divided into static and mobile, direct or indirect protection.

Of all these means, those endowed with the power of mobile direct protection are the most secure, for not only does direct protection nullify a blow at any given spot, but, if it be endowed with mobility, it can be carried, like the carapace of a tortoise, from place to place.

For long this means of protection has been used at sea, and during the Great War it was reintroduced on land in the form of the tank or armoured caterpillar car.

Throughout the history of war there has existed a prolonged conflict between direct protection and movement in order to develop offensive power. Hitting was essential; but was it more economical to protect the hitter or to enable him to move? The result of this conflict was the establishment of two main types of soldiers—the heavily and the lightly protected. Thus we find: heavy and light infantry, heavy and light cavalry, and heavy and light artillery. Whenever a just balance has been maintained between protection, offensive power, and mobility, tactics have flourished, and whenever the balance has been upset, by one or the other becoming paramount or absent, the art of war has either stood still or retrogressed.

A recent revolution of movement, introduced during the present century, which has already influenced protection to a high degree and will increasingly continue to do so for some time to come, is the power of flight, and, if the aeroplane has not already induced us to review the whole of our existing military organization, it will certainly compel us to do so in the near future.

In the past land warfare has been based on one- and two-dimensional movement; the first having normally been used for strategical and administrative purposes, and the second for tactical manœuvres and battle-lines. The second has protected the first by drawing defensive, perpendicular fronts across the strategical and administrative lines of communication, or by enabling troops to take up a position on the flanks of them, and so threaten any attempt on the part of the enemy to occupy them. These are the grand-tactical aspects of direct and indirect protection, and they have been decisively weakened by the present-day power of gaining three-dimensional movement by aircraft, which now enable areas to be attacked as well as fronts.

### 6. The Military Objects and Objectives

In chapter iv. I examined the various objects of war—the national, ethical, economic, and political objects; but I did not include in that chapter the military objects, because, before these can be fully understood, it is, in my opinion, necessary, not only to understand the nature of the military instrument, but to grasp thoroughly the character of the various forces in war. I have now examined these forces, and, as the objectives in war are physical, I will include the examination of this subject in the present chapter.

The military object may be expressed in the one word "conquest," which presupposes victory in one form or another, and by conquest I understand that condition of success which will admit of a government imposing its will on the enemy's nation, and so attaining the execution of its policy. Conquest may also be considered as the grand strategical military idea, and victory the grand tactical military means. Conquest demands the occupation of the enemy's country, and victory the destruction, or disintegration, of his military power, and, as I have already noted, hitherto, on account of the enemy's physical resistance, destruction—especially physical—has monopolized the soldier's mind until it has become the end of war. This is an illogical outlook, since the true political object is to secure a better peace—a securer peace, true, but also a more prosperous and contented peace. Security, prosperity, and liberty rest on certain factors. If these factors are their necessary foundations during peacetime, then in war they must not be destroyed, or if injury to them is unavoidable, it must as far as possible be restricted, and it never can be restricted or avoided if soldiers consider that the main object of war is destruction. It is not, for conquest should aim, not at devastating the enemy's land and decimating his people, but at establishing a condition which will permit of one government imposing its will on another at the minimum ethical, economic, and military cost to both sides, and to the world as a whole.

The reader may remember that in chapter v. I quoted Lloyd as saying that "an army is a machine composed of several parts"—of strength, agility, and universality. Here, I think, we find the germ—even if Lloyd did not fully grasp it—of a fundamental truth. Accepting these terms, I will substitute their forms for their natures. For strength I will write "organization," for agility "tactics," and for universality "strategy." The organic object in war is obviously endurance—for the side which can endure the longer is the side which is going to win; the object of tactics is to attain secure activity—that is, protected offensive power; and of strategy, secure mobility—that is, protected movement. If a general can move where he likes he has attained full freedom of movement, and if he can do what he likes then, equally, has he attained full freedom of action. Both these conditions are obviously ideals, and not realities, since no general can possibly be omnipotent. Yet the nearer he approaches to these ideal states the more economically will he be able to carry out the military object.

Diagrammatically, the relationships of strategy and tactics to force may be shown as follows:

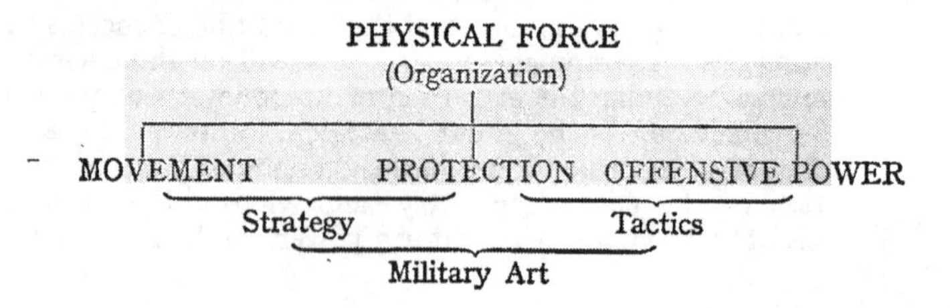

Strategy and tactics cannot be separated; not only are they linked together by administration, which maintains organization, but they are so closely related that unless they interfuse and combine, military art must suffer. In themselves they are abstracts, combined they are a practical reality. One may be paper and the other may be pencil, but art is in the picture drawn; for art is to be sought in the mental and moral forces of the commander and his men, expressing themselves through physical means.

We thus obtain a trinity in which the stable base is organization, the active base is tactics, and the co-operative base is strategy. The sides of these three bases set together form what may be called the triangle of art, and in this triangle the will of the general rules.

If the student accepts these views, then it follows that the object of strategy is to disintegrate the enemy's power of co-operation, and that of tactics is to destroy his activity. The first is attained by placing troops in such a position that the enemy is unable to exert freedom of movement, and is compelled to move according to the will of his enemy. The second is attained by using troops in such a manner that the enemy's freedom of action is restricted, and he is compelled to protect himself in place of hitting out. The first is only attained through the second; and the second is only economically attained through the first; and both, as they are attained, disintegrate the enemy's organization; and as this organization weakens his stability is reduced; and, when sufficiently reduced, the result is victory, and, when totally reduced, it is conquest.

To turn now to the objectives. In chapter vi., when examining the mental sphere of war, I stated that the grand tactical object in war is the destruction of the enemy's plan, and that the decisive point of attack is the will of the enemy's commander. As the base of grand tactics is grand strategy, so is its cutting-edge strategy and tactics, for which no better word than art exists to express the combination of the two. Physical force must be expended in battle, consequently the general, when in a strategic mood, aims at so distributing his force that he may, when the clash takes place, be able to concentrate a superiority of force at and against an objective which will enable him to accomplish his plan and frustrate the enemy from doing likewise. As no army for long can endure unless its system of maintenance remains intact, the strategical objective is the rear of the enemy's army, his supply depôts, communications, and railheads, etc. If these are threatened, then, in place of carrying out his plan, the enemy's commander will be compelled to abandon it and fight for their security, and, until he has secured them, his plan will remain in abeyance.

As I shall return to this subject when I examine and elaborate the principles of war, I will turn to the tactical objectives which, I consider, are not so well understood. Here once again I will quote Clausewitz ; he writes :

> The overthrow of the enemy is the aim of war, destruction of the hostile military forces the means, both in attack and defence. By the destruction of the enemy's military force the defensive is led on to the offensive, the offensive is led by it to the conquest of territory. Territory is, therefore, the object of the attack ; but that need not be a whole country, it may be confined to a part, a province, a strip of country, a fortress. All these things may have a substantial value from their political importance in treating for peace, whether they are retained or exchanged.[1]

And again,

> If a battalion is ordered to drive the enemy from a rising ground, or a bridge, etc., then properly the occupation of any such locality is the real object, the destruction of the enemy's armed forces which takes place only the means or secondary matter. If the enemy can be driven away merely by a demonstration, the object is attained all the same ; but this hill or bridge is, in point of fact, only required as a means of increasing the gross amount of loss inflicted on the enemy's armed force.[2]

This, I think, is a true statement. A position is not in itself an objective to be gained, but only so in relationship to the ultimate object. The seizing of a position may be a means of defeating an enemy, or the defeat of the enemy may be the means of occupying a position ; they are, in fact, relative objectives ; and the second has, in my opinion, not been fully understood, for to defeat an enemy is a complex problem, and not a simple one, as I will now show by means of an example.

A plan of campaign demands a definite object which should never be lost sight of, and this object, in its turn, demands a series of moves each demanding an objective of its own.

The grand-tactical object is to destroy the enemy's plan, and its objective is the peaceful occupation of the enemy's country, which demands the overthrow of the enemy's military power. I will take as my example a type of battle familiar to all soldiers, namely, a trench-to-trench attack, such as was again and again attempted during the first three years of the Great War.

The problem is as follows :

It is our intention to destroy the enemy's plan, the strength of which is based on his power of command and supply, which is protected by several systems of trenches and by artillery and

[1] *On War*, vol. iii., p. 6. [2] *Ibid.*, vol. i., p. 38.

infantry. These trenches must be pierced in order to defeat the enemy's field-army, but in themselves they form no serious obstacles, unless defended by weapons.

There are many of these weapons. Which one is the most vital to the maintenance of their strength? The gun; because the gun forms the base from which rifles and machine-guns operate.

We must attempt, therefore, first to master the enemy's artillery, for, when it is mastered, we shall then, by means of our artillery and infantry, be able more economically to attack his infantry, who, having been deprived of their base of action, have been weakened by a loss of security.

If a house is to be rapidly demolished, we do not attack it from the roof downwards, but at its base—its foundations and lower walls. The roof of a 1916 army was its infantry; its lower walls its artillery; its foundations its command. At this time its foundations could not be attacked directly; the enemy's artillery constituted, therefore, the primary objective.

These guns may, however, be placed between two definite, defended zones, in which case, even if they are captured, other defences will have to be pierced before we can attack the enemy's field-army and system of command. This does not alter the primary necessity of destroying him, but only makes the piercing of the enemy's last line of defences our secondary objective.

To attain both primary and secondary objectives, a series of subsidiary objectives may have to be gained, and possibly also in order to weaken the enemy at the point of attack, it may be necessary to institute certain subordinate tactical operations, which can only be considered of value if they reduce the enemy's fighting power at the decisive point of attack to a greater extent than our own.

From what I have now said can be charted out in tabular form the whole series of battle objectives:

*Grand-Tactical Object*

The destruction of the enemy's plan.

| | |
|---|---|
| *Main Tactical Object* | *Subordinate Tactical Object* |
| To exhaust the enemy's reserves and defeat his field-army. | To induce the enemy to withdraw troops from the point of attack. |
| *Primary Tactical Objective* | *Secondary Tactical Objective* |
| The enemy's artillery. | The enemy's last line of defence. |
| *Subsidiary Tactical Objectives* | *Subsidiary Tactical Objectives* |
| Positions leading to the enemy's guns. | Positions leading to the last line of defence. |

The above example is only an example and nothing more, for each attack, according to the conditions it is likely to be confronted by, will demand individual consideration. The point I have attempted to make clear is this: that every army has an organization, and that the most vital part of the organization becomes the primary objective—the bull's-eye of the target. Armies, like animals, vary in mind and body; some have small brains and large bodies; others have small bodies and large brains; others possess a thick hide; others require large quantities of food; thus I could go on multiplying these characteristics. All possess a variety of limitations; it is the most pronounced of these limitations which we should attack; consequently, though the grand-tactical object remains the same, the nature of the objectives to be attacked vary directly with the nature of the military organization of the enemy's forces and the position they occupy.

### 7. Strategical Formations

From the objective I will now turn back to the instrument, which is an organization possessed of mental, moral, and physical force; and I will examine, not strategy and tactics, which, conjoint, largely constitute the art of war, but the forms of their application.

Strategy mainly consists in combining movements, and security of movement not only depends on local protection, but on the strategical distribution of the forces in the field.

Movement is not only conditioned by the plan adopted, but by the form of the object moved. In war the will of the commander formulates the plan and the strategical formation used is the shape or form of the military projectile. The secret of all economic military formations is that they must possess harmony of offensive and defensive power through movement. Movement in its broadest sense being what I will call "locomobility"—that is, freedom to move in all directions without unnecessary loss of energy or time.

In warfare in which supply is governed by a one-dimensional means of movement locomobility is most difficult to attain. As these are the wars which at present face us, I will first of all outline the main strategical formations of armies as we know them to-day, and when I have done this I will turn to a mechanically propelled army and note how cross-country movement will influence formations.

As the main tactical problem in battle is to give blows without receiving them, the aim of strategy is to place a body of men in such a position that it can most economically solve this problem. The solution is to be sought in the adoption of a formation which will allow of the most rapid approach culminating in the most rapid deployment; for formations must be extended in order that the troops may make the fullest use of their weapons. "Columns," writes Napier, "are the soul of military operations; in them is the victory, and in them is safety to be found after a

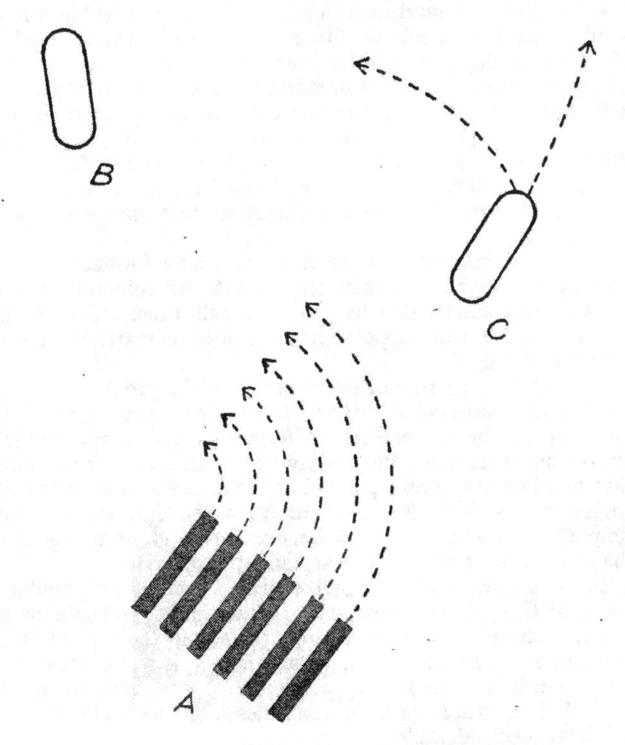


DIAGRAM 2.—COLUMNS IN PARALLEL ORDER

defeat. The secret consists in knowing when and where to extend the front." In other words, to deploy at the right time and the right place is the true foundation of the battle, and, as long as armies cannot move extended, even if it were desirable that they should, columns will have to be employed. I will now examine this question.

### 8. The Column Formation

The simplest form of column is a formation of men in Indian file. On a road, according to its width, this formation is normally stiffened to a column of threes, fours, or eights.

A hundred thousand men in fours, at one yard between fours, would constitute a column fifteen miles long. There would be, therefore, a day's march between its van and rear. If these hundred thousand men are organised as six divisions of all arms, with transport, the length of the column will be approximately five times as great, i.e. seventy-five miles. It would take, therefore, five days for it to pass a given point. Marching at fifteen miles a day, it possesses good mobility, but its locomobility —that is, its power to move at right-angles to its line of advance— is negligible.

As such a column is a most cumbersome formation, I will split up this gigantic human serpent into six columns, and will place these columns side by side and call them Army A (see diagram 2). I will suppose that this army is marching towards a hostile force—C.

Leaving the question of reserves out of the problem, it makes no difference whether A intends to envelop or to penetrate C, for there can be but one march formation which will permit of all A's units striking the enemy together. This formation is that of a line of columns parellel to the enemy's front or at right-angles to his flank (see diagram 2). This formation is very simple, A being in a position either to converge or diverge from the axis of his advance as his plan matures.

Suppose now that a second hostile force, B, is introduced, and that C, by closing inwards or falling back, renders a change of direction on the part of A imperative. Is deployment in line of columns applicable? It certainly is not, for, to change direction towards B, A must order a wheel to the left, and, though the inner division will have but a few miles to go, the outer divisions will have a considerable number.

### 9. The Formation of the Echeloned Line of Columns

Is there no other formation which will enable A to march against C, and, if necessary, rapidly change direction towards B? Yes, there is the echeloned line of columns, on occasion made use of by Gustavus Adolphus. The formation of the echeloned line of columns (see diagram 3) enables A rapidly to engage C with his entire force, and equally rapidly to change direction towards

B, if such a change is demanded. Thus, if the marching front of the six divisions is fourteen to eighteen miles—that is, about two and a half to three and a half miles between divisions—and the depth echeloned back from the head of the leading division to the

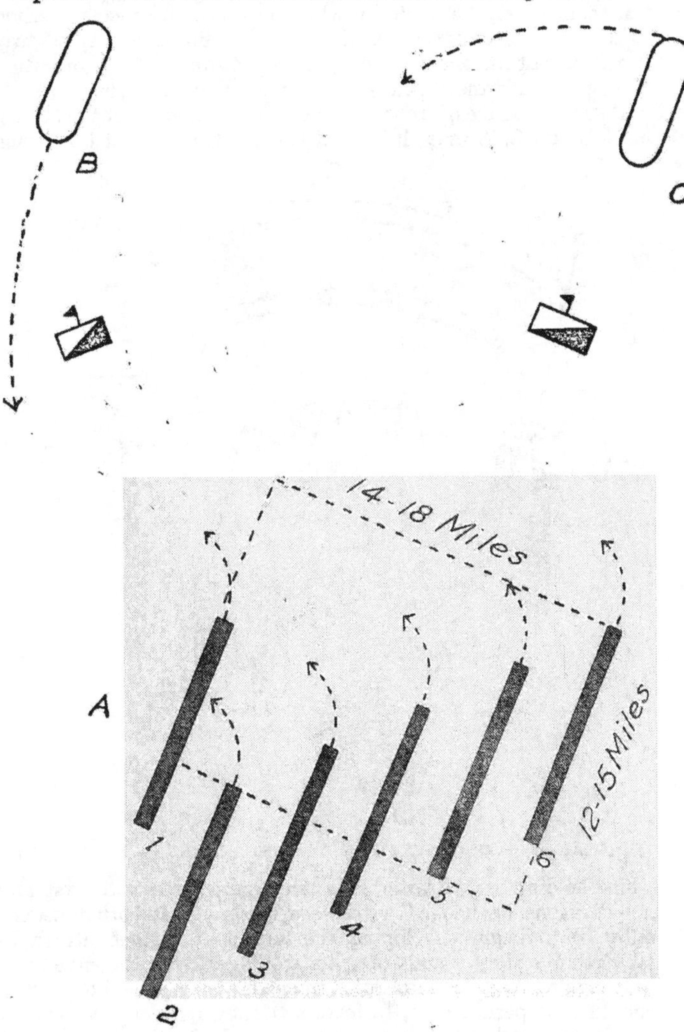


DIAGRAM 3.—COLUMNS IN ECHELON

head of the rearmost be from twelve to fifteen miles, then, if a change of direction from C to B becomes imperative, this change can immediately be made by wheeling the head of each division to the left. The division on the exposed flank should be slightly in advance of the one next to it, in order to allow of the formation of a general advanced guard to cover the change of front.

If such a change of front is impossible, on account of the closing in of B and C, A may, if he still thinks fit, carry out his attack

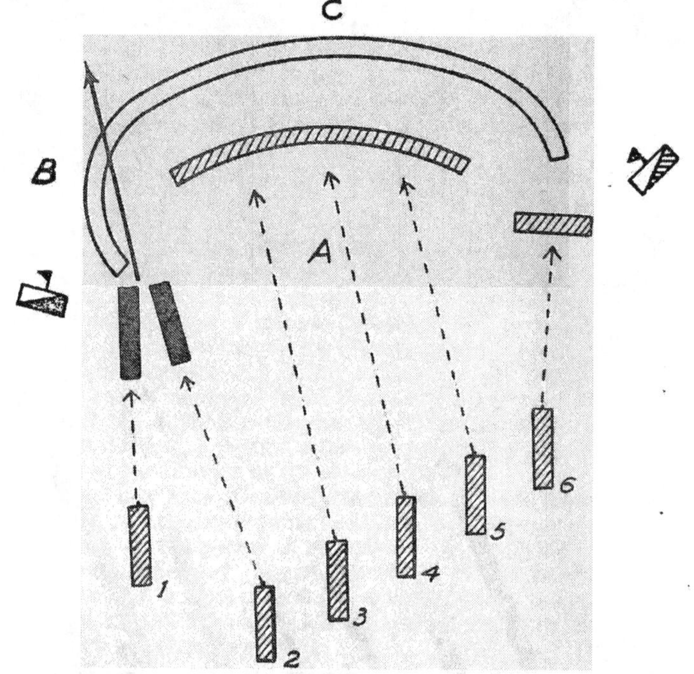


DIAGRAM 4.—CONCENTRATION AGAINST AN ENVELOPING FLANK

whilst holding back B with his cavalry, supported by the 1st and 2nd divisions, or engage C with the 3rd, 4th, 5th, and 6th divisions, allow B to begin enveloping this attack, and then attack in strength B's right flank—in other words, envelop the enveloper.

If, again, B and C unite prior to encounter, A would do better, should time permit of it, to form a triangular lozenge somewhat similar to Marshal Bugeaud's triangle at the battle of Isly (1844) against the now-converging semicircular line, and either hold back its wings as they begin to clinch and penetrate its centre or hold back its centre and destroy its wings by taking them in enfilade (see diagram 4).

Supposing, now, that A detaches one cavalry and one infantry division to operate against B, whilst with the remainder of his force he attacks C, I will examine what factors, outside march formations, will affect his deployment.

Directly a commander knows where his enemy is and when he will meet him he can no longer delay, his plan of action must be

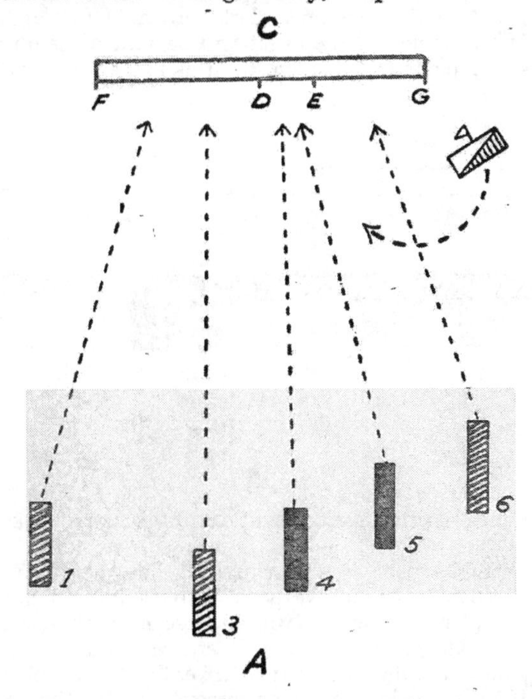


DIAGRAM 5.—CONCENTRATION AGAINST THE CENTRE

formulated, zones of attack must be allotted, the frontages of these zones depending on the probable intensity of the fighting which is likely to take place in each, as well as their relative tactical importance and natural strength. Where is the decisive blow to be struck? This is the keystone of every deployment. If this question cannot be settled before severe fighting takes place, zones of approximately equal size must be allotted to a certain number of units, whilst other units are kept back to reinforce any such zone wherein a decisive advantage is being gained. This will mean that the whole force will not strike together; a separation will take place between the holding and the decisive attacks, which is undesirable. Can this defect be obviated? Certainly, by apportioning zones of action to each unit, the frontages of which are in proportion to their tasks. Thus, suppose that in diagram 5 the area DE offers the main tactical advantage, then the 4th and 5th divisions might be directed against DE, whilst the 2nd and the 3rd hold FD, and the 6th EG. When deployed, the effect will be that of depth opposite the decisive point (see diagram 6); this point being, not necessarily where the enemy is in least strength, but where A can develop the fullest power of all his weapons combined and simultaneously.

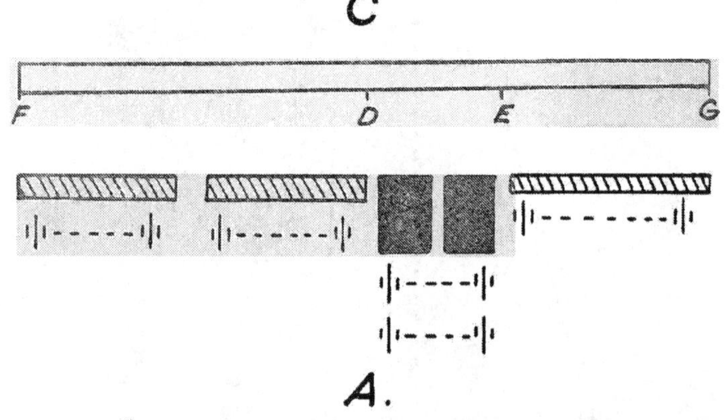


DIAGRAM 6.—ARTILLERY CONCENTRATION AGAINST THE CENTRE

If such a point is found, I will suppose near the enemy's left flank, well and good; the only difference is that FDE will be held by three divisions, whilst two deliver the decisive blow against EG. If such a point cannot be discovered, and time permit of it, an artificially weak point may be created by causing C to weaken one of his flanks, for example, the right, by a threatening envelopment by means of the 2nd division, whilst the 3rd and 4th converge on the weakened section FD (see diagram 7).

In the above formations and movements it should be noted that the security of A's army does not depend on detachments or a general advanced guard, but on ability to attack in bulk and at the shortest possible notice. Co-operation is based on unity of action.

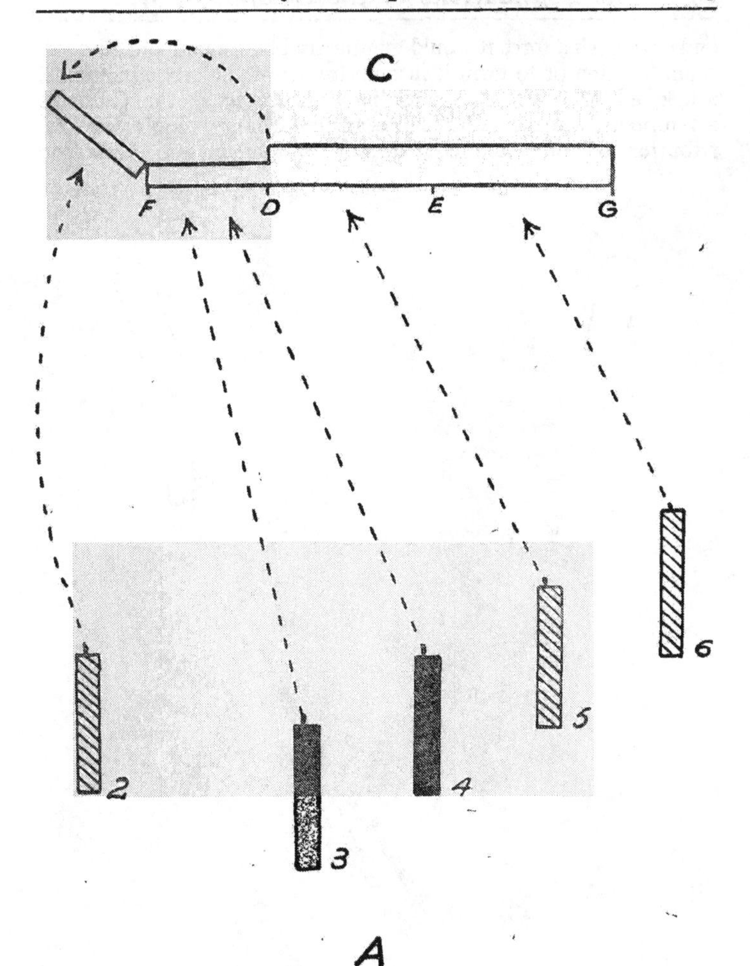


DIAGRAM 7.—CONCENTRATION AGAINST A WEAKENED FLANK

### 10. THE LOZENGE FORMATION

The echeloned column formation is an army formation, and in my example I have dealt with an army of six divisions. If we multiply this number by ten we get an army of sixty divisions, and with such a force it would manifestly be unsound and cumbersome to attempt to form it into an immense phalanx of columns, echeloned or otherwise. This is virtually what the Germans attempted in 1914. With large armies what is required is distribution of force and combination of movements. Napoleon

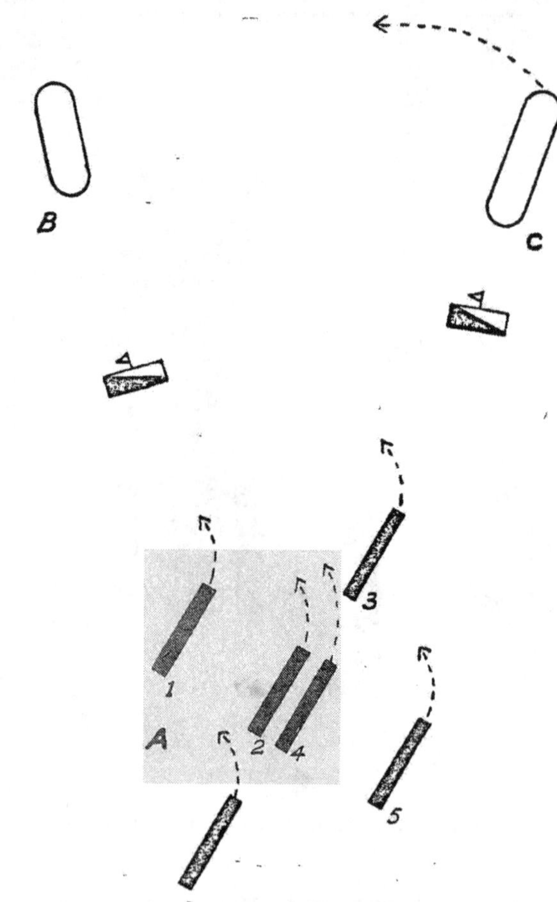


DIAGRAM 8.—THE NAPOLEONIC LOZENGE

understood this well, and he frequently made use of a lozenge formation (see diagram 8). This formation normally consisted of a general advanced guard, two wings, and a central body;

sometimes a rearguard was added. The main advantage of such a distribution is that, whether the enemy is met in front on the right or on the left, he can be engaged by a strong force which will compel him to deploy, and which can hold him until one or more of the other forces are able to concentrate against him. Thus, if the advanced guard first gains contact, the wings can manœuvre towards the enemy's flanks and the central body towards whichever point becomes the decisive point; the rearguard remaining in reserve. If the right or left wings come into contact, an exactly similar series of manœuvres can take place. The great advantage of the lozenge formation is that it combines security and offensive power through movement.

For small forces this formation is well suited to a country in which roads are few and bad. Its main defect is its depth, which scarcely permits of a lozenge of six divisions coming into action on the same day. Consequently in an encounter battle its divisions are liable to become engaged piecemeal. In diagram 8 the 6th division is badly placed for a movement against C, and the 5th division is equally badly placed if a wheel has to be made towards B.

### 11. The Formation of Motorized Armies

I have now outlined the three main strategical formations—paralleled columns, echeloned columns, and lozenge. I have not discussed their tactical advantages and disadvantages. Personally I believe that the defensive power of modern weapons is so great that frontal attacks are no longer reasonable, unless they can be carried out by armoured troops. Further, I believe that, as armour can be carried by machines and, consequently, men can be rendered invulnerable to bullets, it is only rational to suppose that armour will be used. If this is a correct deduction, then the following question arises: If armies are motorized—that is to say, should cavalry and infantry be replaced by tanks and armoured cars—will the above strategical formations prove suitable? Not only will they prove suitable, but much more flexible, for the geometricity of their form, which is most difficult to maintain when roads have to be followed, becomes a fairly simple question over normally open country. Further than this, the restriction imposed by roads being modified, columns, if necessary, can be reduced in depth by broadening their fronts until the maximum breadth of frontage is attained by forming into line. This broadening of their fronts enables them to increase their locomobility by becoming more concentrated. Thus the formation shown in diagram 9 might replace that shown in diagram 3. The *total* frontage is not increased, but the depth of the army is considerably decreased.

Besides this ability to move concentrated, mechanical armies possess the power to move extended. When the position of the enemy is known, this will enable the difficulty noted by Napier—namely " of knowing when and where to extend the front "—to

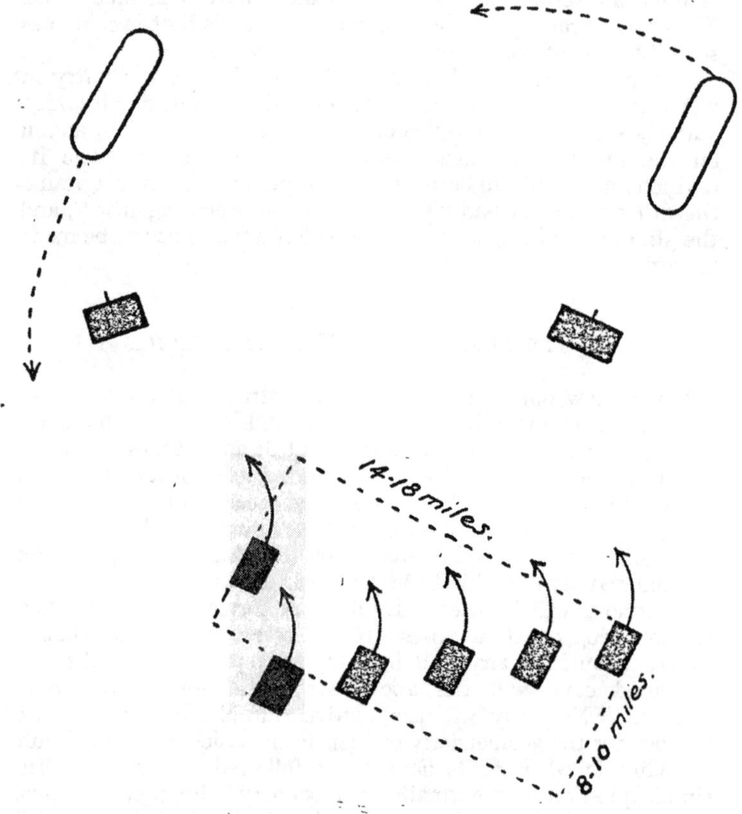


DIAGRAM 9.—MECHANICAL COLUMNS IN ECHELON

be overcome. Normally, however, that is when uncertainty exists as regards the strength of the enemy on the line of his advance, it would appear that mechanical armies will have generally to move concentrated and not extended, but this will not prohibit the use of an extended advanced guard covering the main body. For such an extension it is unlikely that a purely linear formation will be used, but rather that of an arrow-head, strongly reinforced at the apex by capital machines, and flanked by rapidly moving tanks of the destroyer type (see diagram 10).

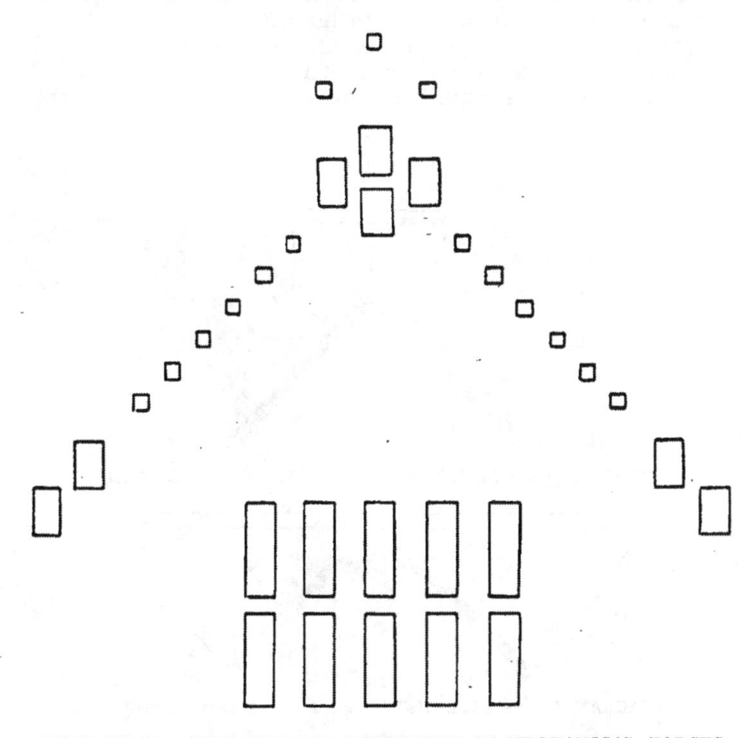

DIAGRAM 10.—THE GENERAL FORMATION OF MECHANICAL FORCES

Behind the advanced guard, the main body can move either in column, line of columns, in lozenge, or in echeloned columns.

If the enemy be met with in strength, the advanced guard can manœuvre for time, or if in weakness, it can forge ahead, driving the apex of the arrow through him, or hold him with the apex and its immediate flanking forces, and swing forward the wings in order to envelop the troops thus held or immobilized. Diagrams 11 and 12 illustrate these two manœuvres.

## 12. THE THREEFOLD ORDER OF TACTICAL ACTION

I will now turn to tactical action, which is developed from strategical formation and distribution, and I will descend to minor tactics.

By strategy an enemy is out-manœuvred; that is, he is placed in a bad position from which to hit out. First it should be remembered that the purpose of tactics is similar to that of strategy, namely to carry out the intention of the commander—his plan. The instrument is not only the troops but the

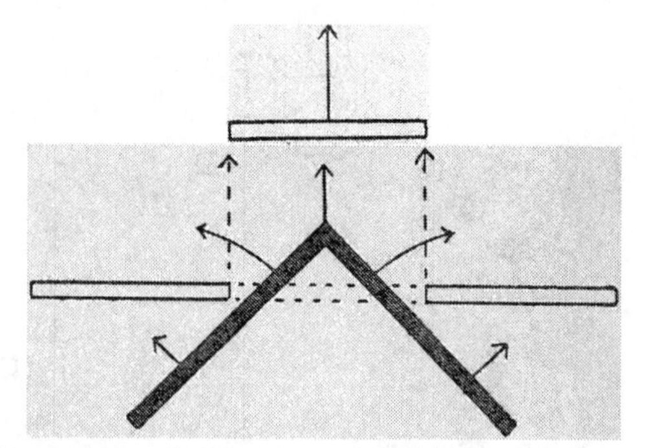

DIAGRAM 11.—PENETRATION BY A MECHANICAL FORCE

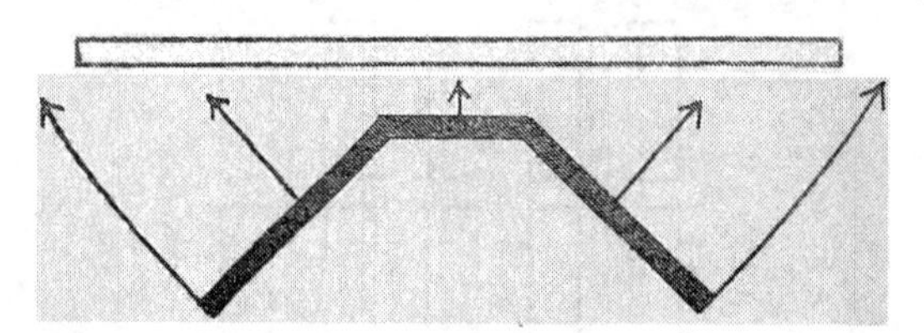

DIAGRAM 12.—OUTFLANKING BY A MECHANICAL FORCE

*organization* of the troops. Organization *must* be maintained. Further movement *must* be maintained, or at least the power to move must exist when the commander desires to move. We here get as our battle problem the maintenance of a moving *organized* body of men. This body must be able to move, and it must remain organized. The enemy is attempting to stop this movement, not only by killing and wounding our men, but *by destroying their organization.* We must, therefore, protect our men and their organization, and we do so to a great extent through offensive action. By hitting we reduce the chances of being hit.

Tactical action may, therefore, be defined as: protected *organized* movement through offensive action.

To accomplish this we require three orders of troops. Troops which will protect the attackers, troops which can attack, and troops which can pursue. These three orders remain fundamental, and to pull their full weight they must co-operate—that is, work together to attain a common object.

In a present-day army these orders are represented by artillery, infantry, and cavalry; and the reason why in the last great war a decision was so long delayed was due to:

(i.) The immobility of artillery.
(ii.) The defensive strength of infantry.
(iii.) The offensive weakness of cavalry.

The number of guns employed and the enormous supply of ammunition required tied artillery down to definite areas, and as intensity of fire had to be maintained, and guns cannot fire when in movement, the result was that when they had to move the attack virtually had to be suspended.

The defensive power of infantry and the lack of ability on the part of cavalry to pursue needs no accentuation.

What we have got to do now is to think in the terms of the elements of war and make good the above deficiencies. Thus, artillery must be endowed with a higher power of movement. Infantry must be endowed with higher offensive power, and cavalry must be more highly protected.

I have laid down three orders of troops from the major point of view, now I will examine them from the minor—the tactical organization and co-operation of the attackers themselves.

According to the accepted theory of war, the true attackers are the infantry. They attack from the base supplied them by the protective troops—the gunners—and on defeating the enemy's infantry, theoretically, they form a base for cavalry action. If, from the major point of view, three orders of troops are necessary, so also are they necessary from the minor. Consequently an infantry platoon should be a threefold organization, and it virtually is one. To prove this I will first divide the platoon into two equal parts, a forward body and a reserve—the left and right fists of a boxer. Both consist of two weapons—a protective weapon, the Lewis gun, and an offensive weapon, the rifle. The object of the forward division is to deprive the enemy of power to move, so that the reserve division may *move* forward and destroy him. The reserve may assist the forward body by protective fire, but, in any case, the Lewis-gun section of the forward body should protect the advance of the rifle section.

Thus we find, in miniature, the tactics of an army repeating themselves in the platoon. The forward Lewis gun is the field artillery, the forward rifle section the infantry, and the reserve is the cavalry and horse artillery. But, whilst theoretically the cavalry in pursuit can move faster than infantry in flight, in the platoon battle the reserve cannot do so. Consequently, whilst in the main battle the object of the infantry is to disorganize the enemy's infantry so that the cavalry can pursue, in the platoon battle the object of the forward division is to fix or hold its antagonist until the reserve division can move forward and

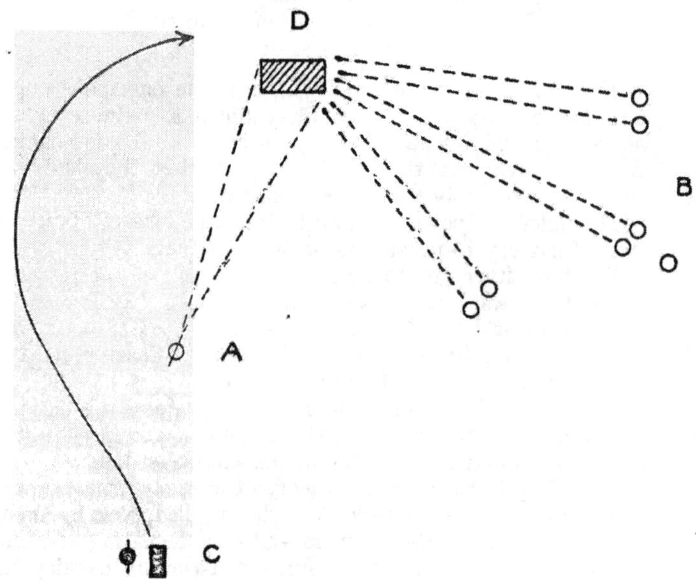


DIAGRAM 13.—THE PLATOON ATTACK

disorganize him. Each time such a disorganization is effected the enemy's battle-body sustains a scratch. In the infantry attack as conceived to-day an antagonist is scratched to pieces.

The diagram (No. 13) shows what I mean. D is the enemy; A is the forward Lewis-gun section; and B the forward rifle section; C is the reserve. Under the protective fire of A, B manœuvres, and through offensive action fixes D. When once D is fixed, C makes the fullest use of movement to manœuvre into a position from which D can be annihilated or compelled to surrender.

Even in so small an action as this we see the close interplay between the three physical elements of war, and, through them, back to the three elements of force. Stability, activity, and co-operation (mobility) demand three types of weapons; these demand three types of soldiers; and these soldiers express their combined action in a threefold order of tactics, namely to protect, to fix, and to destroy or paralyse.

Again we get a close relationship between strategy and tactics. The position occupied by A is first of all tactical—that is, offensive; secondly it is strategical—to cover the movement of B. B's movement is strategical, then tactical; and so also is C's. If strategy and tactics cannot be separated in the platoon, neither can they be separated in an army. Even if our force comprises three men, one should act protectively, one offensively, and the third in a mobile manner; even if only one man, he should protect himself with one fist, hit out with the other, and move by leg-power; and one man is our ultimate model, for one man is our military molecule.

### 13. The Study of the Physical Sphere

In the history of war the physical sphere of force has undoubtedly attracted the greatest attention, as it is the most tangible of the three, yet its study has been alchemical, since system has been lacking, and the result has been, and still is, that, when physical organization has proved itself defective, a remedy is sought for by making demands on the *moral* of the soldier. To strike a comparison: if an engine is the physical means at our disposal, and the engine-driver the moral, then, when the engine refuses to move, in place of examining it and discovering the cause, we say to the driver: "Get out of your cab and push it."

To discover the defects, and, consequently, the improvements in the physical sphere, the physical elements of war are our surest guide; and, if a pass-book will enable a banker to ascertain how a client lives, the forms these elements take in an army will enable a student to discover the mental calibre of its general and higher command. If we see that an army is content with what it has got, this will tell us one thing; if its heads are seeking for higher protective, mobile, and offensive power, then another. In the past evolution has been slow; since science has been backward; but to-day science is leaping ahead, and each leap potentizes the physical sphere, which becomes big with possibilities, so big that it has become not only conceivable but practical for a new weapon to be invented which may give the army equipped with it so great an advantage that nothing can withstand it.

If we value our *moral* as something worth preserving, and the moral school of war mainly looks upon it as cash—something to be spent—then we must never slacken our endeavours to increase physical force in its three forms, since we do not fight with *moral,* but with weapons. *Moral* sustains fighting power, but it does not deal blows.

What armies are to-day doing so? For one of these armies we shall one day have to meet. The mere addition of new weapons and means of movement and protection must not delude us into supposing that an army is guided by progress, for the " test " of progress is *tactical idea.* How are they being used? This is the question. The answer is to be sought in the training manuals and on the manœuvre grounds. Here we can learn how they are being used, and then, possessed with this information, we should turn to the weapons and means and ascertain their powers and limitations. Does tactical theory express them? If it does, then we learn that an army is thinking scientifically; if not, then that its command is composed of alchemists. This is a tremendous and decisive discovery to make.

Next we should examine the military structure of organization. Does it admit the true tactical values of the means being expressed, and does it permit of a co-ordination of tactical structure and maintenance, and is it easily controllable?

To be controllable and maintainable it must be simple. Is it simple or complex? Is it growing like the body of a man, or like an amorphous polypus: that is, is each new means accentuating the power of the elements of war by correlation, or by mere addition? If by addition, then we are faced by a monster, and monsters are seldom to be feared.

As the power of each weapon is limited, so also is the force of an organization limited. What are its limitations, and how can they be overcome. These are a very few of the many questions we should set ourselves to answer, and so prepare ourselves for the next war, not merely by studying history, but by examining the existing organization of all armies, including our own.

Then in war we are faced by another series of questions. What is the object, the idea, in the head of our antagonist? Examine his objectives, his strategy and tactics, and at once a hypothesis can be formulated which will link matter to mind, the outer to the inner, and supply us with an answer. Watch this answer, compare it with facts, amend it, recast it, and, little by little, we creep into the very brain of our enemy and see him as he sees us, and learn his strength and his weakness. Thus, by grasping the essential characteristics of the physical sphere, can we learn to understand the nature of the mental and moral spheres, and act accordingly. The physical sphere is, in fact, the alphabet of war,

# CHAPTER IX

## THE CONDITIONS OF WAR

Perfect uniformity produces no change; all change arise from some difference, from some alteration of balance of conditions.
—G. GORE.

A choice of difficulties seems a necessary condition of human affairs.—ARCHBISHOP WHATELY.

### 1. A THREEFOLD ORDER OF CONDITIONS

I HAVE now dealt with the instrument of war and its forces, and more particularly with the military instrument, and though in the main I have had the idea of an army before me, I am of opinion that in principle the examination I have now concluded can be equally well applied to a navy or to an air force. From these forces I will now turn to those which change and modify them, and the causes of these changes I will call the conditions of war, which include every possible cause which can produce an effect in the instrument.

In chapter iii. I stated that the universe is known as a space of three dimensions, which manifests to us in terms of time and force, and that knowledge, faith, and belief are the varying relationships between these three conditions and the mind. In war these three conditions surround us as completely as they do in peace, but as our minds are concentrated on a single and highly specialized problem, namely the waging of a war, they assume relatively a military aspect, and, in order to distinguish them from their more general forms, I will call them military space, military force, and military time. We thus obtain two trinities—the general and the special—the first relative to life as a whole, and the second to war as a special problem. Thus graphically these two trinities can be shown as in diagram 14.

In the first triangle, each change in space, force, and time influences man; in the second triangle, each change in military space, force, and time influences the military instrument. In the first case, unless the mind of man can grasp the nature of the changes which are bombarding him his life will be the resultant of trial and error; if he can, then of knowledge. Knowledge will tell him that these changes can assist him, resist him, and

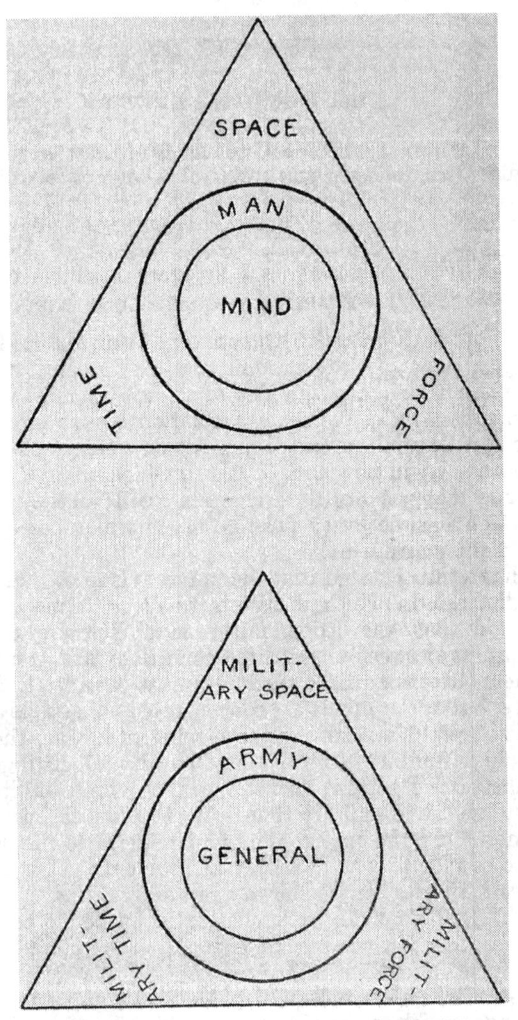


DIAGRAM 14.—THE MILITARY TRINITY

transform him, his transformation being, in fact, the relationship between assistance which is active and resistance which is stable. In the second case it is the same, for if the mind of the general can understand the conditions which are influencing his army he will be in a position to avoid, resist, or turn these conditions to his advantage, and thus strengthen his own army and weaken his adversary's. In fact, he will be able to transform the fighting power of both. Action resulting from such knowledge may be termed scientific action, in contradistinction to action which does not, which is alchemical action.

The number of the conditions of war may be considered as infinite, consequently this rules out of all possibility the power of one mind grasping them as a whole. To overcome this difficulty —or, rather, to limit it—a general is assisted by a staff, the main duty of which is to examine the conditions of war and to deduce their influences. It is in this important work that the scientific method will assist us. I will illustrate this by a quotation :

Mr. F. W. Westaway writes : " . . . with even the closest attention, our observations may be entirely incorrect. Any one of our organs of sense is easily deceived, a fact which enables the magician to make his living. Then it is seldom that we see the whole of any event that occurs : a cab and a bicycle collide, and half a dozen 'witnesses,' all perfectly honest, may—probably will—give accounts which differ materially and may be mutually destructive. It is always difficult to keep fact and influence distinctly apart. In the middle of the night we 'hear a dog bark in the street.' But really all that we hear is a noise ; that the noise comes from a dog, and that the dog is in the street, are inferences, and the inferences may be wrong. For instance, a boy may be imitating a dog ; and everybody knows how easily the ear is deceived in regard to the direction of sound. It is almost impossible to separate what we perceive from what we infer ; and we certainly cannot obtain a sure base of facts by rejecting all inferences and judgments of our own, for in all facts such inferences and judgments form an unavoidable element." [1]

For a moment I will pursue this problem of noises. Suppose, for some reason or another, we wish to specialize in noises ; then we must examine all possible noises in turn, and, though we may never be able to acquire a complete knowledge of all noises, we shall obtain knowledge of a considerable number. Then, when a noise occurs, especially a common noise, such as a dog barking, we shall be able to infer its cause with far greater accuracy, and sometimes even the reason of its cause ; in fact, by a scientific study of noises we shall become experts in the subject.

[1] *Scientific Method*, F. W. Westaway, p. 195.

Turning now from noises to the conditions of war. Though in their totality, they are infinite, or presumably so, we know that those which are constantly repeating themselves are limited in number. From a close study of military history and the psychology of nations we shall be able to deduce by far the greater number of general conditions; and from a careful study of our own and the enemy's instruments of war and the characteristics of the probable theatres of war we shall further be able to deduce a large number of special conditions.

This will give us a sound foundation to build accurate inferences on, but we must not rest here, for we must ascertain what the probable influence of these conditions will be on ourselves—the mind of the general and his army. It is here that the elements of war can render us true assistance as checks to our judgments. We know that conditions will influence all the three spheres of forces, and that, as each of these spheres contains three elements, one or more of these elements will be affected. Which are most or least affected, or will be so? Once we have answered this question, though we may not have arrived at the truth, our decision is more likely to be true than if founded on mere guess-work.

In brief, every change in the conditions of war produces a change in the forces of the military instrument and transforms it, whether we like it or not. What are these transformations? They are changes in the elements: in the mental sphere—changes in reason, imagination, and will; in the moral sphere—changes in fear, courage, and *moral*; and in the physical sphere—changes in offensive power, protective power, and mobility. Many of these conditions are occult; that is, they are hidden until they manifest; but by far the greater number of the common ones are obvious, such as: a courageous man will fight better than a coward; two men should exert greater force than one; a protected man is not so vulnerable as an unprotected; a concise order is more easily understood than an involved one; night operations are more susceptible to panic and disorder than those carried out in daylight; a surprise attack is more economical than an expected one; a hilly country is less easy to cross than a prairie; an infantry man is useless against a tank; a horse cannot carry as heavy a load as a lorry, etc., etc.

There are several hundreds of these common conditions which recur in every war, and which in the past have had to be relearnt in every war, because the soldier will not, or cannot, think scientifically. Commanding an army, organized, I will suppose, for war on the plains, a general enters a mountainous region and is annihilated, and he cannot understand why. Simply because he has not foreseen the influence of conditions—in this case of physical geography—on the forces and structure of his instrument. In 1755 General Braddock attempted a Horse Guards parade against Red Indians in the Monongahela forests, and was crushingly defeated. Just before he died he murmured: "Another time we shall know how to deal with them." But why wait for next time? In 1914 we constantly hurled infantry against barbed wire protected by machine-guns; in 1915 we beat a naval gong outside the Dardanelles, and then ordered our soldiers to land; in Mesopotamia we forgot to send out an adequate supply of bandages and surgical instruments; and so on *ad infinitum*; and why? Simply because we would not think in terms of the conditions of war, and discover the influence of these conditions on the instrument. "Pour in sow's blood, that hath eaten her nine farrow; grease that's sweaten from the murderer's gibbet, throw into the flame"—that was our method, and yet we were not so successful as the witches in Macbeth.

## 2. The Conditions of Military Time

The division of the conditions of war into the categories of time, space, and force has at least the advantage of simplicity. Strategically, these categories form the base of all our calculations, and tactically of all our actions, and each may be considered as possessing either an abstract or a concrete mood. I will now very briefly examine these three general categories of conditions from their military aspect.

Time is an all-embracing condition, and in war, more so even than in peace, time must be reckoned in minutes, and not only from a military point of view, but from an economic one as well, since in a war, such as the Great War of 1914–18, every minute of time was costing Great Britain from four to five thousand pounds.

The economy of time becomes, therefore, not only of military but of economic importance; it is never unlimited in its remunerative sense, and its loss can seldom be made good; in fact, of all losses it is the most difficult to compensate. One of the greatest problems in generalship is how to utilize time to the best advantage, and this demands a perfectly organized instrument in which friction, which is the enemy of military time, is reduced to its lowest possible level. To understand the time limitations of one's own side and of the enemy's is to work from the surest of foundations, and if our organization will enable us to move more rapidly than the enemy, then from the start we possess an immense advantage over him, for indirectly this organization will enable us to increase the time at our disposal.

Economy of time first depends on thoroughness of preparations, and secondly on stability of policy. If a nation which is parsimonious during peace-time enters upon war unprepared to wage it, it will either succumb to force of hostile superiority or else will be compelled to pay an enormous premium in order to make good its peace-time deficit. A want of preparedness must detrimentally affect any policy, preconceived or improvised. Without fixity of purpose there can be no military stability, for changes in policy are the most fruitful sources of delay. Besides, economically the cost is stupendous, for every hour lost may be £250,000 thrown away, a little less than the price of the upkeep of two battalions of British infantry for one whole year. Again, if full preparations are made during peace-time, and the war, once it has begun, proves to be totally different in character from the war expected, the greater part of these preparations will have been wasted. Thus we see—and especially so in modern times—that, though the soldier frequently blames the politician for refusing to vote more money for preparations, the politician, if he knew anything of war, might well retort that the money is being withheld, not to stop preparations, but to prevent preparations which will prove useless. If in the next war we are confronted by a mechanicalized army, even if in peace-time we possess ten times the infantry we have, we shall be less well prepared to meet this war than we are to-day, since we shall have squandered millions and millions of pounds.

Time, strategically, is the measurement of military movement; tactically, of muscular and mechanical endurance. Time is, therefore, intimately related to the means of movement, protection, and weapons. These constitute, in fact, the works of the military clock. Time, also, frequently means concentration and economy of force. Thus, if time can be economized, numbers can either be multiplied or reduced, especially if an operation is carried out so rapidly that the enemy is unable to meet it. Superiority of time is so important a factor in war that it frequently becomes the governing condition.

### 3. The Condition of Military Space

The practical application of time is the utilization of space, which strategically and tactically, since the advent of the aeroplane and the submarine, has become three dimensional. Formerly space, from its military aspect, was two dimensional as regards tactics and one dimensional as regards supply. The addition of a second dimension to supply, by means of the cross-country tractor, and of a third dimension to tactics, by means of the aeroplane, both petrol-driven machines, has ushered in a new military epoch.

Military spaces can no longer be reckoned in terms of areas which are actually occupied by armies, or which separate them. Formerly, armies had frontages of attack with a tactical space between them, which was contended for, and the importance of which could be calculated by appreciating the value of the tactical features in relation to the enemy's intentions and communications. To-day all this is changing, since armies are rapidly becoming three-dimensional organizations. Spaces have grown to include, not merely battlefields or theatres of war, but whole countries, and so much so is this the case, that it is quite possible to visualize an army holding at bay another, whilst its aircraft are destroying the hostile communications and bases and so paralysing enemy action.

Spaces are now no longer definitely restricted by rivers, deserts, or mountain ranges, for to a great extent these space walls have been surmounted by the aeroplane, which renders impotent so many natural and artificial obstacles, and so frees military time of its greatest spendthrifts.

Spaces include the three mediums of movement, namely water, air, and earth. At present each requires a special means of movement; thus, water requires ships; air, aeroplanes; and land, wheeled or tracked vehicles. Consequently the present restrictions of space require three differently constituted fighting forces—navies, air forces, and armies. Should in the future, however, a means of movement be discovered which will enable one machine to combine the powers of present-day sea, land, and air machines, space, in the military sense, will become universal; its walls will have ceased to exist. The storming of the bastions of space is the greatest military problem of the future.

From purely a land point of view, military space, though measured in miles, kilometres, etc., should generally be considered with reference to resistance; just as time should be considered with reference to the probable intentions of the enemy. Thus, in an entrenched battle our line of trenches may be separated from the enemy's by a hundred yards, yet if the intervening space be well wired it may take longer to cross it successfully than one hundred miles of open country. Space, like time, in its military aspect, must always be equated with force, and the conditions which assist, resist, and transform force.

### 4. The Conditions of Military Force

I have dealt at such length with military force as a compound of nine elements operating in three closely related spheres that it is not necessary for me to return to this subject; in place, I will examine the conditions which influence the interplay between the two forces represented by the two military instruments—the enemy's fighting forces and our own. In each sphere we find two sub-categories of conditions—the natural and the artificial. For instance, in the mental sphere we have the genius of the commanders, which may be considered as natural mental conditions. We also have the machinery of information, which is an artificial condition. In the moral sphere of force we have racial character, which is natural and training, which is artificial; and in the physical sphere we have weapons, means of protection and of movement, which are artificial, and ground, weather, and geographical conditions, which are natural.

It is obviously impossible for me, within the limits of a single volume, to examine in any detail this host of conditions; consequently I will restrict myself to a few general remarks on each of the three categories.

### 5. The Mental Conditions of War

The mental conditions of war, though shared between the general and his men alike, are of supreme importance to the former, just as the physical conditions are to the latter. The general is the centre of greatest responsibility; and command, as I have shown, is as much a matter of self-government as of the government of others; it is he, in fact, who fights, and he fights with his brain; and if he wins, he reaps the glory of victory, and if he loses, then the ignominy of defeat. Responsibility in war is the heaviest load any man can carry; to suggest is easy, to do is indeed hard.

The conditions of war are appraised by the general, or at least they should be, for his staff is only a sorting-machine which in no way can relieve him of his responsibility to decide. His plan must in every way be his own plan, whether he has devised it himself or borrowed it from another, and, be it remembered, there is nothing wrong in borrowing; much has to be borrowed in war, and history offers us innumerable suggestions. What is wrong is merely to copy without reference to conditions; equally is it wrong to initiate without this reference; conditions are, therefore, the spirit-level and plummet-line of a general's plan.

A plan of war is always confronted by another plan, however vague it may be, and between the two plans lie the conditions of war which, to the opposing generals, are very largely mental in character. These conditions may be considered as the unknown $x$ in an equation, and on the plus and minus values of this $x$ will the actions of both sides depend.

Thus, diagrammatically, this may be shown as follows:

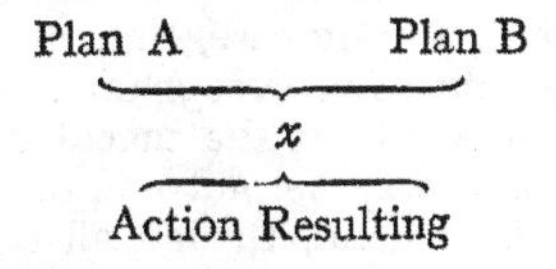


What, now, does $x$ represent?

It represents to a very large extent the influence of the enemy, of the instrument, and of the general's native *moral*.

As to the value of the first, this is almost self-evident; as the pressure the enemy brings to bear rapidly becomes felt, and is frequently understood in the physical and moral spheres of war. Yet in the mental sphere its understanding is often vague, since only the greatest generals, and then more through intuition than reason, have grasped the mental conditions which surround their adversary. Is he a free agent, or a mere political tool? Is he an artist, or a mere mechanician? Does he believe in the doctrines promulgated in his army, or does he not? Is he the slave, or master, of his staff? And, above all, is he a man who has studied war scientifically, or alchemically?

In 1914 I much doubt whether any single general possessed more than a passing knowledge of his enemy's or ally's commanders. Did General Joffre really understand General French; or General French, did he understand General von Moltke? I make bold to say that not one out of ten generals in the British army had ever heard of either von Moltke or Joffre; yet they were training their men to fight the Germans and to co-operate with the French. They thought, if they thought at all—and through no fault of their own, but because of the system in which they worked—of the physical side of the approaching war, to the complete exclusion of the mental. Like drill instructors, they taught their men to aim and to fire, or they watched others teaching them, but they paid no attention—or very few of them did—to the mentality of their enemy's command, and they never drilled themselves into understanding that, when it came to battle, it was going to be a fight between their ideas and their opponents' ideas, and not merely between their men and the enemy's. To us these mental conditions were all but a complete blank, because we had never troubled to study character, and to-day we still have no machinery wherewith to do so.

The mental conditions of the instrument—the will of the Government, the will of the staff, and the will of the soldiers—all act and react on the will of the general. Is he proof against these influences, and can he maintain his own equilibrium? Consider his surroundings. His staff may or may not be of his own choosing; in any case, they are all very human; some are self-seekers, some are sycophants, some are full of ideas, and some are mere grit in the machine; yet, however efficient or inefficient they may be, not one of them can share the responsibility of the general, though all can influence his will, unless this will be of steel. If he is a judge of character, and if he possesses a deep knowledge of human nature, the general will understand the mental conditions which surround him; mere stubbornness will not do this. To refuse to listen to advice is not a token of strength, but of stupidity, a vice only second to that of weakness. It is through an intelligent grasp of his surroundings, the mental conditions which form the instrument of his work, that a general succeeds in freeing his will from obstruction. If his men murmur, and he knows why they murmur, he can act rightly; and if his staff suggest, and he knows the character and mental calibre of each member of his staff, then will he know the psychological value of each suggestion. Finally, he must understand his own moral force and work within its limits. This of all his problems is the hardest to solve.

As regards the men he commands, they must understand the use of the physical elements, and not merely possess skill in their use. A condition suddenly manifests—it may be a clump of trees seen from a rise in the ground, or an unexpected trench, or an unlooked-for machine-gun, or one of the ten thousand minor conditions which incessantly ripple over the battlefield. Does every man understand simultaneously what each of these conditions means, and its influence on the situation at the moment? For unless they do understand them their skill will to a large extent be wasted. Not only must they understand them from their point of view, but from that of the enemy, so that they may equate the two series of factors resulting and arrive at a true decision. And, when they have decided, will they act? This depends on the condition of their *moral*, and generally this is a question for their leaders to decide.

## 6. The Moral Conditions of War

As I have examined at some length the moral sphere of war, I will deal very briefly with the moral conditions, which in general terms must be understood by the commander, and in detail by the leaders of the men themselves.

It must be remembered that all conditions—or very nearly all—influence the soldier morally by stimulating either his courage or his fear; for, whilst some affect war materially, such as roads for supply and the influence of gravity on the flight of projectiles, thousands directly stimulate the instincts of the soldier, and through his instincts his mind, and through his mind his actions.

Examining this question from a very general point of view, the various moral conditions of war may be divided into three main groups, namely:

(i.) Those which are general; that is, those which influence men individually and collectively.
(ii.) Those which more especially influence the individual.
(iii.) And those which more particularly influence a mass of soldiers as a homogeneous crowd.

The following are examples of these groups:

(i.) *General Conditions:* Safety, comfort, fatigue, catchwords, loyalty, honour, faith, hatred, love, admiration, cheerfulness, etc.
(ii.) *Individual Conditions:* Knowledge, leadership, command, skill, determination, reason, endurance, courage, self-confidence, stubbornness, sense of duty, etc.
(iii.) *Collective Conditions:* Suggestion, intuition, instinct, superstition, esprit-de-corps, tradition, example, religion, education, patriotism, comradeship, etc.

It is not possible to draw a hard and fast line between these conditions, for they overlap, and I do not propose to analyse them, as each would require a separate chapter. Nevertheless it must be realized that, unless these conditions are understood, it is not possible to apply efficiently the principles of war, and, unless all the conditions which go to build up soldiership have been stabilized prior to the outbreak of a war, a general will not possess a stable vehicle for his will to move in. The process whereby this stability is gained is called training. Training forms the true foundations of battle which, just as war should be a continuation of peace policy, should, in its turn, be a continuation of peace training. War is, in fact, the examiner of all our work.

For this to be possible it will at once be seen that training must be based on:

(i.) The permanent characteristics of man.
(ii.) The permanent characteristics of war.
(iii.) The probable conditions in which the next war will be fought.

These conditions must be foreseen, and, as war is an evolution of civilization, the tendencies of civilization must be discovered. On the correct forecasting of the nature of the next war will depend the continuity of peace training when war breaks out, under the changed form of battle tactics.

There is really no great difficulty, if application be made, to foresee, with a fair degree of accuracy, the tendencies towards improvement in weapon design, etc.; but, unless the psychology of war has been carefully studied, there is a distinct difficulty to forecast the moral conditions, new weapons, etc., will give rise to on the battlefield. Thus, for instance, a tank can undoubtedly assist an infantryman to capture a machine-gun, but will this increase the courage of the infantryman? Not necessarily; for, in place of stimulating his courage, the fact that the tank is invulnerable to machine-gun fire will throw him back on his reason and imagination, and he will say: "This machine is quite capable of dealing with the machine-gun; why should I risk, therefore, my life by following it closely? I will wait until the tank has destroyed the enemy, and then I will advance and occupy the position." This is common sense, and we must understand such conditions as these, for otherwise we may, during peace-time, when the instincts are not aroused (because of the absence of danger), determine on tactics which demand close co-operation between tanks and infantry, and then, during war-time, we may discover that the infantry *will not* closely co-operate, and our tactics break down, because they are not harmonized with the moral conditions created by the tank in this special case—the infantry attack. There are hundreds of these problems which face us to-day.

## 7. The Physical Conditions of War

The physical conditions of war permit of a definite distinction being made between the artificial and the natural. In the former category we have the two opposing instruments, each comprising weapons, and means of protection, of movement, and of supply, of repair and of transportation; each creating strategical, tactical, and administrative conditions, which affect mutual changes in force and in organization. In the latter we have geography, topography, and climate, and also in this category may be counted communications, political centres, and industrial areas, for, though these are not natural conditions, they lie outside the province of military control.

To examine with any completeness the various physical conditions of war would demand, not only a book, but a series of books; obviously, therefore, I cannot do more than accentuate their importance. Lloyd considered that the theatre of operations is " the great and sole book of war." This, within the limitations of the physical sphere of war, is a correct statement. During war we have little time to read this book, and, unless we have closely studied it before the outbreak of war, the application of our means will be profoundly restricted.

In this study the civil sciences can help us, and are progressively becoming, not mere handmaids of the soldier, but his closely collaborating partners. To render this collaboration possible it is most necessary for the soldier to realize that, though he is the expert authority on the application of means, the scientist is the expert authority on their creation. The problem which faces the soldier is how to adapt action to circumstances. Circumstances are the conditions of war; action is the use of the military instrument. The instrument cannot be omnipotent; consequently its powers, however formidable, must be limited. What are these limitations, and how will conditions affect them? This question can only be answered by discovering what the nature of the conditions is. This is still a military problem. We know, or should know, with fair accuracy the conditions of the last war, the nearest war to any war which to-day confronts us; but, however full our knowledge may be of this war, we must never forget that a war to-day, or a war to-morrow, even if fought over the same theatre as the last war, will not be the same, even if military science and art has stood completely still during the intervening period.

The reason for this is that, however lethargic the soldier may be during peace-time, it is during peace, much more so than in war, that the struggle for scientific knowledge and industrial survival is acutest. Each new discovery, each new invention, by modifying the forces of peace modifies the force of war. The soldier must understand these modifications, because in the next war they will confront him as actual conditions. The next war is his supreme problem. An examination of national characteristics and international politics, of peace treaties, of frontiers, of economic influences, and of ethical ideals, will enable him approximately to arrive at the date of the next war and to define its theatre. Suppose all these tendencies point to a war against Russia between the years 1935 and 1940, here, then, is a sound hypothetical base to work on. What will be the conditions of this war? To arrive at an answer we must analyse the existing world situation and discover its political and scientific tendencies. Once these tendencies have been discovered, we must work synthetically, and, guided by our hypothesis, project these discoveries into the future. Here the political philosopher and the scientist can help us. We can ask questions; they can give us provisional answers. With these in our mind we can first compare the limitations of our existing military instrument with the most probable conditions which will confront us in a war with Russia between the years 1935–40. Secondly, we can fall back on our provisional answers and modify the powers of the instrument. We shall then arrive at the conception of a hypothetical instrument, varying from the existing one in characteristics *a*, *b*, *c*, *d*, etc. Suppose *a* represents a gas-proof tank, *b* an aeroplane with a radius of action of one thousand miles, and *c* a persistent gas which will remain potent for one month, then we can turn to the scientist and say, Here are three problems to solve; solve them!

We now have got a clear idea of what we want; that is to say, we have an object in our heads and an objective as our goal; what must we next do? Not merely wait for the scientist to give us what we want, but to think out first the tactical use of these new inventions, and, when our tactical ideas are clear, secondly, to change gradually the structure of the military instrument so that it may become an efficient vehicle for the full powers of these new weapons to express themselves.

But suppose we have made a political miscalculation. Suppose in 1937 we are at war with Germany and not with Russia. Conditions will certainly be different, though perhaps not radically so. This is a possibility we must not overlook; therefore we must take each possible, even if not probable, war in turn and arrive at its conditions, and through these at the changes in our military instrument. These must be compared and correlated. Those which are found to be contradictory or mutually incompatible we must examine in the light of our imagination, guided by our hypothesis of the most probable type of war, and to those which only disagree in detail we must apply our reason and so discover an answer.

To-day no army in the world possesses a general staff which can think in the terms I have outlined, yet one day some nation, I am convinced, will possess one, since it is but common sense that it should possess one, for its cost is insignificant. In our own case, the money we yearly spend on the Bermuda garrison would, I imagine, go a long way to pay for its establishment.[1]

### 8. The Conditions of Ground

I propose now to turn to the natural physical conditions and examine only three, namely ground,[2] weather, and communications, and merely as examples, for the number of important natural conditions is very great.

The practical expression of space is ground, in which to-day are to be sought the main obstacles to movement in land warfare. Ground may be divided into three main types:

(i.) Mountainous country.
(ii.) Undulating country.
(iii.) Plain lands.

The nature of each of these types is normally governed by water. If water be abundant, the following conditions are generally met with:

(i.) In mountainous country: swift rivers, unsuitable as communications, and wooded valleys.

(ii.) In undulating country: large rivers as great thoroughfares, and towns and scattered villages.

(iii.) In plain lands: an extensive network of rivers and towns and scattered farmsteads; or few rivers and consequently desert regions.

The influence of water on the soil itself and the influence of soil on civilization are most marked. Thus, where the rainfall is normal, flat countries will usually possess a high water-level, and undulating countries a low one. This frequently means that in flat countries the inhabitants will live in scattered houses and farms, and that in undulating countries they will live in villages, the houses of which are congregated round a few communal wells.

[1] In 1925 the cost of the garrison of Bermuda was £119,300, £28,800 being spent on Royal Artillery, the men of which were costing £327 a head. During the same year the garrison of Mauritius was costing £34,700, of which £23,100 was being spent on Royal Artillery.

[2] Clausewitz has many interesting remarks to make on ground. See his *On War*, vol. ii., pp. 120, 121, 127, 128, and 238, and vol. iii., p. 183.

From a tactical point of view this will mean that flat countries are usually good defensive areas, and undulating ones good offensive areas, as the latter will offer fewer natural and artificial obstacles. The meshes between the knots—the villages—will be bigger than between the farms, consequently movement will be facilitated.

The influence of ground on military organization is considerable, and one of the greatest difficulties of the army organizer is to fashion an organization which will be sufficiently elastic to prove suitable in all natures of country. This in the past has proved almost as difficult as squaring the circle, but to-day the solution to this problem would appear to be rendered possible by the aeroplane and the cross-country car which, by replacing muscular endurance by mechanical energy, will to a great extent annul the differences of ground, by rendering movement over, or on, the various types more feasible.

### 9. The Conditions of Weather

Weather is not only to a great extent a controller of the condition of ground, but also of movement. It is scarcely necessary to point out the influence of heat and cold on the human body, or the effect of rain, fog, and frost on tactical and administrative mobility; but it is necessary to appreciate the moral effect of weather and climate, for in the past stupendous mistakes have resulted through deficiency in this appreciation.

Human nature, as I pointed out in chapter vi., is continually influenced by its surroundings. These surroundings vary considerably, not only in the theatre of war, but throughout the armies operating in it. I will illustrate what I mean by an example.

A battle is being fought on a hot day. The temperature on the battlefield is 100° in the shade; consequently the soldiers are directly influenced by the heat. A few miles behind the front the headquarter staff officers are seated in a house in which the temperature is 80°. They may be working under electric fans; they are not carrying 50lbs. on their backs, and are probably in their shirt-sleeves. If they are thirsty, they can call for a drink. The conditions in which the battle is being controlled and those in which it is being fought are diametrically opposite.

Unless the headquarter staff have intimate experience of the conditions surrounding the fighters, two types of battle are likely to be waged—the first between the brains of the army and the enemy, in which case this action will be rendered impotent on account of the muscles being unable to execute the commands of the brains; and the second between the muscles and the enemy, which battle will be disorganized, not so much through the enemy's opposition as through the receipt of orders which are impossible to carry out.

It will be said: "But it is the duty of the headquarter staff to keep in intimate touch with the fighting troops." Of course it is; but there is a great difference between laying down a duty and carrying it out, especially during war-time.

Instead of placing the staff in similar conditions to those prevalent on the battlefield it is the first duty of the military designer to create an army which will enable the soldier on the battlefield to be placed in conditions resembling, so far as possible, those the staff are situated in. The object is not, therefore, to accentuate the discomfort of the whole, but to minimize the discomfort of the part, and in the above example this means that the temperature of the muscles must be brought down to that of the brains.

At first thought this might appear to be an impossible problem; on second thought it will be realized that it is not so if the soldier is provided with a means of movement which will enable him to bring with him on to the battlefield such comforts as will square the difference. To-day the cross-country tractor, or the tank, will enable him to go into action with an electric fan and a whisky and soda. Further, the tank will force the headquarter staff to get into similar machines in order to keep up with the fighting troops, so that the equation will be still more completely solved.

### 10. The Conditions of Communications

Closely related to ground and influenced by weather are communications, which are even more important administratively than they are strategically, for the supply system of an army may be compared to the blood of the human body—it constitutes, so to speak, the vital fluid which keeps the whole organization alive. With masses of men the maintenance of supply unavoidably becomes of greater importance than tactics. The army has got to live in order to fight, and, as living is most difficult, supply consequently becomes its primary problem and fighting its secondary problem.

Communications may be divided into three categories:

(i.) Strategical communications.
(ii.) Administrative communications.
(iii.) Tactical communications.

Each or all may include means of movement by air, sea, or land, and land communications depend, in civilized warfare, on roads, railways, rivers, and canals, all of which are in nature one-dimensional. Ever since the introduction of the wheeled cart this linear nature of communications has been one of the controlling conditions in land warfare.

The restrictions which the one-dimensional nature of land communications has imposed on the strategical, administrative, and tactical movement of armies have been stupendous, the difficulties steadily increasing with the growth of armies, in spite of the invention of the locomotive and the lorry.

During 1914–18 this limitation was the predominant factor of the war; it was no longer a question of manœuvring to protect communications, but of increasing communications in order to move. Road-capacity was the controlling condition, and so it is likely to be in every future war, unless roads can be dispensed with and land communications made in nature two-dimensional by means of cross-country traction. This means, supplemented by the three-dimensional power of the aeroplane, will revolutionize totally the administrative organization of armies.

## II. The Dual Power of Conditions

In the first section of this chapter I stated that every condition of war possessed a dual power, namely, of assistance and of resistance to the instrument of war. For instance, if an army is organized for war in open country a mountainous region is apt to resist its organization, and an open one to assist it. Physical conditions, such as woods, hills, defiles, rivers, swamps, etc., can be used, therefore, to accentuate or lower the power of the instrument, just as various materials can accentuate or lower the power of a tool. If we want to bore a hole through a piece of steel we use a drill suitable for this purpose, and not a bradawl. To a general, the conditions of war are wood or steel, and generalship largely consists in compelling an enemy to bore holes through the latter whilst we are boring holes through the former. To do so, a general must possess knowledge of the conditions of war. He must know all he can before war is declared, and discover all he can during its progress; consequently observation, information, and reconnaissance are essential factors in war.

Information must be collected, evalued, and correlated with the forces of the instrument, and action must be planned to assist in this correlation. If we turn to the history of war, we shall discover that a commander has three means at his disposal in order to deal with a condition:

(i.) He may avoid it.
(ii.) He may force it aside.
(iii.) And he may turn it to his advantage.

The third course, which masters the difficulty, is manifestly the best, and it is the one which even a superficial study of military history will show us was employed by all the great captains of war; it was, in fact, the keystone of their success. To turn conditions, however adverse, to advantage, is, in fact, the test of good generalship, and to do so we must understand the relationship between pressure and resistance. This brings me to the law of economy of force.

# CHAPTER X

## THE LAW OF ECONOMY OF FORCE

All Nature is but art unknown to thee;
All Chance Direction which thou canst not see;
All Discord, Harmony not understood;
All partial evil, universal Good;
And, spite of pride, in erring Reason's spite,
One truth is clear, whatever is, is right.
—POPE.

### 1. THE UNIFORMITY OF FORCE

I HAVE now dealt with the forces of war, and have shown that changes in the external forces—namely, the conditions of war—produce changes in the internal forces of the instrument of war, and modify its structure, and influence its maintenance and control. The question now arises, can any general laws, principles, or rules be formulated whereby we may judge the change wrought by any set of conditions on the forces of the instrument, and, through them, on our intention? If war is a science, or is reduced to a science, as a consequence such laws, principles, and rules are axiomatic, for science lays bare the nature of relationships and discovers the reasons upon which they are based. There must be, therefore, certain laws or principles of war, just as there are laws of chemistry, of physics, and of psychology.

I have already stated in chapter ii. that war is not an exact science, and by this I do not mean that fundamentally exactness does not exist—for it must exist in all sciences—but that the human brain is too limited in its power to devise a complete science of war that exactness does not appear to be a possible attainment. Truth must be exact, for inexactness and truthfulness are contradictory terms. Science, which aims at discovering truth, must consequently aim at exactness, even if only an approximate exactness is attainable. We realize this very definitely when we study history. We cannot hope to succeed if we only apply the scientific method, because, as one writer says: "History is a philosophy of transcendental ideals beyond the scope of science, and depends, also, upon emotional literary inspiration to enforce its lessons." [1] In medicine it is likewise,

[1] *The Lessons of History*, C. S. Leavenworth, p. 16.

only an approximate exactness can be attained, because each patient differs psychologically, yet, if we know the causes and natures of the various diseases, we shall be in a better position to cure than if we do not. Meteorology is a science, yet an inexact one, and so also is finance. This does not deter meteorologists and financiers from proceeding with their work ; in fact, it is an incentive for them to do so.

Inexactness, like chance and ignorance, is a quality of the human brain ; it does not exist in Nature. From general observation, our assumption is that Nature is exact, that not a leaf falls to the ground which, within the conditions in which it fell, could possibly have fallen in any other way than it did, or at any other moment. Outside the mind of man, all things are governed by the law of uniformity, and man himself is also governed by this law, but with this difference, that whilst a stone cannot disobey this law, man can, and is meted out punishment in proportion to his disobedience.

I have shown that the forces of war and those of life generally are synonymous. For the time being I will set aside, therefore, the nature of war as a psychological as well as a physical struggle, and look upon it purely as force, and, from this restricted aspect, attempt to establish a general principle which governs the changes in force. Then, when once this principle has been discovered, I intend to make it my base of action and to return to the problem of war, and from it deduce a series of subordinate principles which will assist us to control and expend military force economically —that is, according to the nature of the relationships between the instrument and the changing conditions which surround it.

As my datum point I intend to adopt the system outlined by Herbert Spencer in his *First Principles*. In chapter xii. of this book, a chapter of recapitulation, he says :

> The play of forces is essentially the same in principle throughout the whole region explored by our intelligence ; and though, varying infinitely in their proportions and combinations, they work out results everywhere more or less different, and often seeming to have no kinship, yet they cannot but be among the results of a fundamental community.[1]

Thus the forces of war must take their place in this grand group of forces, and, as Spencer is the philosopher with whom

[1] *First Principles*, H. Spencer (fifth edition), p. 276. In the study of war the military student will find that some knowledge of philosophy is of the greatest assistance. If the student has little time at his disposal for this study, I can recommend, besides Spencer's *First Principles*, the works of David Hume, four volumes, and, if these be found too long, then Thomas Huxley's essay on "Hume," which is a masterpiece of clear thinking. To read Huxley alone is a valuable training.

I am best acquainted—a philosopher who has attempted to work out a synthesis which embraces all sciences—I intend to make him my master and guide, and, in place of paraphrasing and condensing what he says, I will quote from him in full, leaving it to the student, should he wish to amplify these quotations, to turn to the book and earn reward by studying it.

## 2. The Law of Force

In Nature "all is causal, nothing is casual."[1] This is our starting-point, the bed-rock upon which the philosophy of science erects certain universal inferences which are called laws,[2] and which are the abstract descriptions of qualities of facts that are of a general nature, such as "The Uniformity of Nature"; "The Indestructibility of Matter"; "The Continuity of Motion"; "The Persistence of Force"; "The Persistence of Relations among Forces," etc.

Force, according to Herbert Spencer, is the "ultimate of ultimates." To him, space, time, matter, and motion are either built up of or abstracted from experiences of force. He writes: "Thus all . . . modes of consciousness are derivable from experiences of Force; but experiences of Force are not derivable from anything else. Indeed, it needs but to remember that consciousness consists of changes, to see that the ultimate datum of consciousness must be that of which change is the manifestation; and that thus the force by which we ourselves produce changes, and which serve to symbolize the cause of changes in general, is the final disclosure of analysis."[3]

To us force manifests as matter moving in space, the duration of the movement being time. Consciousness of movement is only possible since it possesses two modes, one actual and the other potential. The first occupies space, and the second, which possesses power to effect changes, is generally called energy.

Changes in energy are governed by the law of causation, which

[1] *Logic*, Welton, vol. ii., p. 165.

[2] "A general law or truth is arrived at by detecting a constant or uniformity amongst variables. . . . Rules are based upon laws, and laws are based upon facts. . . . General laws do not rule, they are not causes, nor effects, nor actual things, but brief statements of relations of things" (*The Scientific Basis of Morality*, G. Gore, pp. 1, 15). "A law of Nature is not a uniformity which must be obeyed by all objects, but merely a uniformity which is, as a matter of fact, obeyed by those objects which have come under our observation" (*Principles of Science*, S. Jevons).

[3] *First Principles*, H. Spencer, pp. 169–70.

is a law of motion.[1] Causes by their motion produce effects; thus, if I pull the trigger of a loaded rifle the whole sequence of events which follows originates from muscular motion on the trigger, the primary cause of the sequence.[2] Whether the final cause of change is the workings of a single force, or the conflict of two forces, cannot be determined; but the manifestation of change is the co-existence of pressure and tension, or, as Herbert Spencer says: "Matter cannot be conceived except as manifesting forces of attraction and repulsion,"[3] and "probably this conception of antagonistic forces is originally derived from the antagonism of our flexor and extenser muscles." These two manifestations of force are "our symbols of reality," and from them there result certain laws of direction of all movement. "Where attractive forces alone are concerned, or rather are alone appreciable, movement takes place in the direction of their resultant; which may, in a sense, be called the line of greatest traction. Where repulsive forces alone are concerned, or rather are alone appreciable, movement takes place along their resultant, which is usually known as the line of least resistance. And where both attractive and repulsive forces are concerned, or are appreciable, movement takes place along the resultant of all the tractions and resistances. Strictly speaking, this last is the sole law; since, by the hypothesis, both forces are everywhere in action. . . . Motion then, we may say, always follows the line of greatest traction, or the line of least resistance, or the resultant of the two: bearing in mind that though the last is alone strictly true, the others are in many cases sufficiently near the truth for practical purposes."[4]

[1] "Causation is really the ideal reconstruction of a continuous process of a change in time" (*Appearance and Reality*, Bradley, p. 60). See also *Principles of Logic*, Bradley, pp. 485–8. "Causation acts in such an order that we must first satisfy our bodies by means of food, air, a dwelling, fire, and clothing; then our animal desires, feelings, and emotions; and lastly, our intellect and reason, consequently the last is extensively neglected. Even the determination of human actions by mere desire or feeling is evidence of natural causation; and it is manifest that all education is dependent upon a practical belief in the law of universal causation, otherwise we could not expect any certain effect from personal training" (*The Scientific Basis of Morality*, G. Gore, p. 48).

[2] This sequence can, of course, be carried back further: thus, the finger is pressed because the eye sees an animal, which the mind intends to slay, because hunger demands food, because food is lacking, etc., etc. It would appear that any threat to create a vacuum at once sets the chain of cause and effect vibrating.

[3] Hume states that we know nothing of the feeling we call power except as effort or resistance. Huxley, in his essay on "Hume" (*Collected Essays*, 1897, p. 149), writes: "If I throw a ball, I have a sense of effort which ends when the ball leaves my hand; and if I catch a ball, I have a sense of resistance which comes to an end with the quiescence of the ball. In the former case there is a strong suggestion of something having gone from myself into the ball; in the latter, of something having been received from the ball. Let anyone hold a piece of iron near a strong magnet, and the feeling that the magnet endeavours to pull the iron away in the same manner as he endeavours to pull it in an opposite direction is very strong."

[4] *First Principles*, pp. 224–6.

On account of the interplay between attraction and repulsion, "It further follows from the conditions that the direction of movement can rarely if ever be perfectly straight. For matter in motion to pursue continuously the exact line in which it sets out, the forces of attraction and repulsion must be symmetrically disposed around its path; and the chances against this are infinitely great."[1] Then, a little later on, he writes: "As a step towards unification of knowledge we have now to trace these general laws throughout the various orders of changes which the Cosmos exhibits. We have to note how every motion takes place along the line of greatest traction, of least resistance, or of their resultant: how the setting up of motion along a certain line becomes a cause of its continuance along that line; how, nevertheless, change of relations to external forces always renders this line indirect; and how the degree of its indirectness increases with every addition to the number of influences at work."[2]

Herbert Spencer next examines the operations of these laws in the celestial and terrestrial systems, then in relation to living things, and finally in relation to mind. To summarize his reasoning; he says:

> Supposing the various forces throughout an organism to be previously in equilibrium, then any part which becomes the seat of a further force, added or liberated, must be one from which the force, being resisted by smaller forces around, will initiate motion towards some other part of the organism. If elsewhere in the organism there is a point at which force is being expended, and which so is becoming minus a force which it before had, instead of plus a force which it before had not, and thus is made a point at which the reaction against surrounding forces is diminished, then, manifestly, a motion taking place between the first and the last of these points is a motion along the line of least resistance.[3]

When this motion is frequently repeated, if the channel along which it flows is affected by the discharge, and "if the obstructive action of the tissues traversed involves any reaction upon them, deducting from their obstructive power, then a subsequent motion between these two points will meet with less resistance along this channel than the previous motion met with; and will consequently take this channel still more decidedly. If so, every repetition will still further diminish the resistance offered by this route; and hence will gradually be formed between the two a permanent line of communication, differing greatly from the surrounding tissue in respect of the ease with which force traverses it."[4]

[1] *Ibid.*, p. 227. [2] *Ibid.*, p. 227. [3] *Ibid.*, p. 235. [4] *Ibid.*, p. 236.

From the relation between emotions and actions, Spencer finally turns to volition, and considers an act of will " an incipient discharge along a line which previous experiences have rendered a line of least resistance. And the passing of volition into action is simply a completion of this discharge."[1]

One corollary from this must be noted . . . namely, that the particular set of muscular movements by which any object of desire is reached are movements implying the smallest total of forces to be overcome. As each feeling generates motion along the line of least resistance, it is tolerably clear that a group of feelings, constituting a more or less complex desire, will generate motion along a series of lines of least resistance. That is to say, the desired end will be achieved with the smallest expenditure of effort. Should it be objected that, through want of knowledge or want of skill, a man often pursues the more laborious of two courses, and so overcomes a larger total of opposing forces than was necessary, the reply is, that relatively to his mental state the course he takes is that which presents the fewest difficulties. Though there is another which in the abstract is easier, yet his ignorance of it, or inability to adopt it, is, physically considered, the existence of an insuperable obstacle to the discharge of his energies in that direction. Experience obtained by himself, or communicated by others, has not established in him such channels of nervous communication as are required to make this better course the course of least resistance to him. . . .[2]

Having seen that matter is indestructible, motion continuous, and force persistent—having seen that forces are everywhere undergoing transformation, and that motion, always following the line of least resistance, is invariably rhythmic—it remains to discover the similarly invariable formula expressing the combined consequences of the actions thus separately formulated.

What must be the general character of such a formula? It must be one that specifies the course of the changes undergone by both the matter and the motion. Every transformation implies rearrangement of component parts; and a definition of it, while saying what has happened to the sensible or insensible portions of substance concerned, must also say what has happened to the movements, sensible or insensible, which the rearrangement of parts implies. Further, unless the transformation always goes on in the same way and at the same rate, the formula must specify the conditions under which it commences, ceases, and is reversed.

The law we seek, therefore, must be the law of the *continuous redistribution of matter and motion.*[3]

Spencer then shows that every change undergone by every sensible existence is a change towards integration or disintegration. " But though it is true that every change furthers one or

[1] *Ibid.*, p. 238. [2] *Ibid.*, pp. 238, 239. [3] *Ibid.*, pp. 276, 277.

other of these processes, it is not true that either process is ever wholly unqualified by the other."[1]

Everywhere and to the last, therefore, the change at the moment going on forms a part of one or other of the two processes. While the general history of every aggregate is definable as a change from a diffused imperceptible state to a concentrated perceptible state; every detail of the history is definable as a part of either the one change or the other. This, then, must be that universal law of redistribution of matter and motion, which serves at once to unify the seemingly diverse groups of changes, as well as the entire course of each group.

The process thus everywhere in antagonism, and everywhere gaining now a temporary and now a more or less permanent triumph one over the other, we call Evolution and Dissolution. Evolution under its simplest and most general aspect is the integration of matter and concomitant dissipation of motion, while Dissolution is the absorption of motion and concomitant disintegration of matter.[2]

Here I will leave the philosophy of Herbert Spencer and return to the subject of war.

### 3. Economy of Force

The redistribution of force, such is the ceaseless labour of the universe, a collecting and a dispersing, a mobilization and a demobilization, and perpetual change in unceasing motion, in fact, a war without a victory. Such is the nature of the world as it moves on with cadenced step through endless time and space. Nothing is created, nothing is lost, yet all things are changing, for nothing is standing still, and every change is in accordance to law, until we come to life, and then we find that the supreme problem of all living things is to learn how to obey.

Obedience may be unconscious or conscious; the first leads to evolution through trial and error, the second to progress through rational thought. The first is the common process of the animal world, and to those men who are higher than animals it is the second. To animal existence chance is an omnipotent power, but to the thinking man it is an illusion, for it does not exist, for his reason tells him that omnipotence is law. "War," writes Clausewitz, "is the province of chance. In no sphere of human activity is such a margin to be left for this intruder, because none is so much in constant contact with him on all sides. He increases the uncertainty of every circumstance, and deranges the course of events."[3]

[1] *Ibid.*, p. 283. [2] *Ibid.*, p. 285. [3] *On War*, Clausewitz, vol. i., p. 49.

Clausewitz is only relatively right, right in so far that chance rules when ignorance abounds, and, though we cannot hope to replace ignorance so completely by knowledge that ignorance will vanish, the more we realize that war is the province of law and not of chance the more we shall grow to understand its changes, and, as we understand them, learn how best to economize and expend our force. One author writes :

Untrained man wastes nearly everything with which he has to do, and especially that which is plentiful and cheap—such as water, coal, and food ; he wastes his time, life, health, and opportunities ; he wastes his life largely in idleness or excess of amusement ; his health in selfish excesses ; his opportunities through want of decision and promptitude, and by mistaken conduct ; his mental health by neglecting to acquire wisdom, by filling his mind with trifles, by dwelling upon grievances, or upon irrational "pious" desires. He wastes his physical health and food by eating and drinking to excess, and he wastes time in unnecessary exercise in order to counteract the evil effects of these.[1]

Thus, when we turn to military history, we find that war has mainly been an instrument of waste, because of the ignorance of the soldier. Truly Clausewitz writes : "Every unnecessary expenditure of time, every unnecessary *détour*, is a waste of power, and therefore contrary to the principles of strategy."[2] War is not governed by chance, but by law, and the punishment for disobedience is waste.[3] The rational distribution of force, this is our problem in war.

To Herbert Spencer, force is "the ultimate of ultimates," and to us soldiers so are the forces of war ; not because we want war, but because our *raison d'être* is to expend force in war. Force endures, whatever may be the use made of it ; that is to say, it persists in itself ; but for practical purposes it is limited, for we deal in changes of force, consequently the law of causation governs force in war, which manifests in the form of pressure and tension, and these we call offensive and protective action. As abstract conceptions, they are our "symbols of reality," and, as concrete acts, they are our efforts. Our will moves our muscles, and our muscles enable us to hit and to guard, and by means of hitting and guarding we expend our mental, moral, and physical energy.

If, in its entirety, we could grasp the law of causation, we could then so economize our force that, whatever force might be at our

[1] *The Scientific Basis of Morality*, G. Gore, p. 89.
[2] *On War*, Clausewitz, vol. iii., p. 153.
[3] "It is essential to the idea of *law* that it be attended with a sanction, or, in other words, a penalty or punishment for disobedience" (A. Hamilton, *The Federalist*, p. 210).

disposal, we should expend it at the highest profit. Consequently, if two opponents face each other, and each possesses an identical supply of force, the one who can make his force persist the longest must win, because, as Spencer says, "the desired end will be achieved with the smallest expenditure of force." Therefore, in place of talking of the law of causation, or of the law of persistence of force, as the fundamental law of war, I will call this law the law of economy of force, or the law of economic expenditure of force. The latter term expresses my idea more closely, but as the former appears to me to be more general and scientific, I shall normally make use of it.

### 4. Economy of Mental Force

Spencer, having probed and examined the foundations of knowledge, postulates the law of the continuous redistribution of matter and motion. From this postulate he develops his theory of evolution, and, after examining a great number of facts, he proves his theory to be correct, and to be applicable not only to the subjective world, but to the objective world as well. Thus this theory becomes a law—a living expression of the original postulate.

I have already touched upon this law in the second chapter of this book, in which I explained how evolution works by means of an unceasing process of trial and error. Truth exists only in one form, truth derives its power from economy of force, and trial and error, after endless experiment, arrive at truth by economizing force; perfect economy of force and truth are therefore synonymous.

Darwin, and others, have traced the law of evolution in the physical world. To him it may be summed up as a process of struggle for existence, in which the fittest survive; and fitness not only depends on bulk strength (concentration of force), but on facility of adaption to environment (distribution of force). This law governs us all; and in the vegetable and animal worlds effect follows cause in blind rotation. Man is not blind, for he possesses power to reason. This power I have already examined in chapter vi. and in chapter ii. by means of a quotation I explained, that "if one course of action proves successful and another fails, *there is a reason* for it." By grasping the laws which regulate causes, man can control causes. Reasons express the quality of things, and, if man can understand these qualities, he can learn to use them.

From the law of economy of force we know that there can only be one reason. A cause cannot have various reasons, and if at first the reason appears compound, it is because we do not thoroughly understand it. "Errors," writes Paul Carus, "do not exist in the world of objective facts. Errors are children of the mind. There is neither good nor bad, neither right nor wrong, neither truth nor falsehood, except in mentality. And again: "Truth and error are the privilege of mind."[1]

Do not let this mislead us, for I have just stated that the process of evolution is that of trial and error. Trial and error, as it appears to man, who can reason, and not as it is in Nature, which is swayed by omnipotent cosmic law.

For example, why has a hare got long legs? To escape from the fox and the wild dog! What made its legs long! Thousands of years of snapping and snarling of wild dogs immediately in rear of its tail. The legs grew through a process of trial and error. This is exactly how armies have grown and still grow.

Turn to the racehorse.

Why has the racehorse got long legs? To win the Derby and St. Leger. What made its legs long? A few years of scientific thought and careful selection. Its legs grew through the efforts of man's mind. This is exactly how armies should but do not grow.

In the purely material world there is rigid law; in the physical world there is trial and error, until out of consciousness creeps reason, which applies law to the events and circumstances which surround life.

> The same operations which are active everywhere, separations and combinations [writes Dr. Carus], build up the human frame, and in the human frame also man's mind. Human reason is a structure built up by mind operations; and pure reason is a mental construction of them in abstract purity. The human mind being a part of the world, we find that the law of sameness holds good also for the products of purely mental operations: the same operations yield the same results.

And again:

> Reason is not purely subjective. Reason is objective in nature. Our subjective reason, human reason, or the rationality of our minds grows out of that world-order which we may call the rationality of existence. Human reason is only the reflection of the world-reason; the former is rational only in so far as it agrees with the latter.[2]

The senses enable us to appreciate the effects of causes; reason enables us to discover not only the cause, but the purpose of it—

[1] *Primer of Philosophy*, Paul Carus, pp. 22, 48.
[2] *Ibid.*, pp. 112, 117.

its validity. Reason consists first of " the operations that take place among mental images, secondly it enables us to grasp certain qualities of Reality, and thirdly it is the instrument which enables us methodically and critically to deal with any kind of experience."[1]

> "The facts of experience are specie, and our abstract thoughts are bills which serve to economize the exchange of thought. If the values of our abstractions are not ultimately founded upon the reality of positive facts, they are like cheques or drafts for the payment of which there is no money in the bank."[2]

The reality of positive facts is the goal of the scientific method (the searching for truth methodically), and this method consists, as Mach has observed, in an " economy of thought." It is hence that all economy must proceed. If our thoughts are chaotic, so also will our actions be chaotic; consequently discipline of mind must precede discipline of body, and without the cohesion of these two economy of force cannot be effected.

Throughout the history of war we discover that, in spite of man's ignorance of the science of war, the law of economy of force has been in ceaseless operation. The side which could best economize its force, and which, in consequence, could expend its force more remuneratively, has been the side which has always won. Frequently bulk weight of numbers has won through, and often has it lost. Consequently on first thought, we might be led to suppose that the law I have propounded is no law at all, and that, as God has so often sided with " big battalions," numerical superiority is the surest panacea of victory. But, if we examine history, we shall find that some of the most decisive victories have been won by the numerically weaker side, because it was better led or equipped. From such battles we may deduce the fact that numerical superiority is only a special interpretation of the meaning of strength, and, if this is a correct deduction, then that a science of war is required which will enable us to discover the ingredients of military strength in all its forms. We see, therefore, that military force does not merely depend on numbers, or generalship, or political courage, but on all these requirements and on many others as I have already explained. It is a compound of all activities which can be utilized in war; and a weakness, or deficiency, in any one of these may spell disaster if circumstances favour the enemy.

In war we cannot hope to possess a maximum value of each item of military power, but what we can hope to do is to establish

[1] *Ibid.*, pp. 117–18. [2] *Ibid.*, p. 1.

a science which will enable us to know what these items are, and the nature of the conditions in which they manifest their full values. Then, if certain items are deficient in our military structure, we shall be able to avoid those circumstances in which they will assume predominating values; equally, if we understand conditions, we shall be able to extract the greatest advantages from those items we do possess. It is by knowing what items are present or deficient in our nation and army, and in the enemy's nation and army, and by understanding the conditions of war which stimulate and depress each item, that we shall be able to expend our power profitably, and thereby economize our national power for the pursuits of peace.

### 5. Economy of Moral Force

As the general tendency of man's mind is towards thinking economically—that is, towards discovering the reasons why certain quantities and qualities assist and resist us, so also, in the moral sphere of force, "The fundamental rule of righteousness, that we should do unto others as we would have them do unto us under like circumstances, is evidently based upon the principle of causation, viz., that the same cause always produces the same effect under the same circumstances, for if it could not be depended upon in all cases, the rule based upon it could not be fully trusted."[1] Thus morality in its turn is based on economy of force in the moral sphere.

It may have taken many hundreds of generations to reveal to primitive man (and many are still primitive) that truthfulness, honour, honesty, generosity, gratitude, loyalty, tolerance, and unselfishness, etc., are economical moral qualities—that is to say that they assist human evolution, and that their opposites impede it. At first he may have seen how often a thief or a liar seemed to succeed, whilst an honourable or a truthful man failed; but little by little, as his knowledge expanded, he saw that these apparent exceptions were not contradictions, they did not contradict morality, but were due to some uneconomical condition in the moral system of society, a system which can never be absolutely perfect. It is not because honesty is good and dishonesty is evil that we are honest, but because honesty is essential to salvation, not in the next world, but in the present one. So also with the soldier; trial and error little by little impressed on his mind the economical values of courage, sense of duty, loyalty, obedience, comradeship, self-sacrifice,

[1] *The Scientific Basis of Morality*, G. Gore, p. 2.

patriotism, esprit de corps, etc., and that their opposites undermined moral strength. It was trial and error that showed the way to the mind of man, and revealed to him his power of reason. It supplied him with true facts whereon to build hypotheses, and then it left man to his reason to prove his assumptions. Thus, whether consciously or unconsciously, the law of economy of force has ruled the moral sphere just as it has ruled the mental.

To think rightly is to economize the powers of the brain, and to possess righteous sentiments is to economize the powers of the soul. In both spheres economy of force rules with an iron hand, and punishes every man who refuses to bow to this supreme and all-pervading law.

### 6. Economy of Physical Force

In the physical sphere we see this law in its most manifest form. The whole tendency of work and mechanical progress is towards economizing physical force. At the base of nearly every new invention we find economy written in capital letters. In war this is as visible as in peace. A stone axe economized fist-blows, an iron axe was an economy over the flint axe, the musket over the bow, the rifle over the musket, and so on from the opening of military history to the present day.

To economize man's strength, to economize in life, to economize in numbers, by perfecting the means of war—that is, by rendering them more and more efficient, in spite of imitation, prejudice, ignorance, and stupidity—has been the law of mechanical progress in war, and nothing, outside the whole human race becoming demented, can stay its course. Because a few purblind and talkative humanitarians decided at Washington, a few years back, to abolish chemical warfare, if chemicals are an economical means of waging war, their abolition is about as certain as a dictum to abolish the moon. In the eleventh century Canute understood this full well, yet in the twentieth we find men, who are considered intelligent, misunderstanding it. This certainly shows that the truth-seekers must possess the patience of Job.

To understand what the physical progress of war means, we must apply economy of force to hitting power, to protective power, and to movement. We must not halt here; we must take man and render him skilful in the use of these means according to the various conditions which confront him and are likely to do so.

In training, our first lesson is economy of thought, our second economy of sentiment, and our third economy of physical energy.

Without these lessons, trial and error will continue to be our master; with them, we can make trial and error our slave. Reason is supreme; and any restrictions on freedom of thought during peace-time will sow a crop of tares which will be fully reaped in war. To progress is to economize; to retrogress is to squander; to stand still is to rot.

Thus we see economy of force ruling the three spheres, adapting action to circumstances, and modifying all mental, moral, and physical forces according to the influences of their surroundings. The power of a rifle on a rifle-range may be, $x$ on the battlefield it may be $x-y$. What is $y$? It is all the influences which the conditions of the battle bring to bear on the firer, such as restrictions of view, perturbation of mind, exhaustion of body, and the grip of fear. All these conditions, and many others, influence the firer mentally, morally, and physically. With an army it is the same, and in war, unless the general-in-chief be a supreme genius, a man whose fingers are on the pulse of the battle, a man who can read the innermost meaning of the pulsations of the strife, economy of force, though ever our master, is too abstract a conception to prove a useful guide. Consequently, from this all-controlling law of war, I will attempt to extract certain principles of war, which, having been tested again and again throughout the history of war, have proved themselves true governors of military thought, of sentiment, and of action.

# CHAPTER XI

## THE PRINCIPLES OF WAR

There are principles that make apparent
The images of unapparent things.

—LONGFELLOW.

We extend knowledge by the discovery and accumulation of facts, and we condense it by means of principles, general truths, and laws.—G. GORE.

### 1. THE SEARCH AFTER PRINCIPLES

THE value of principles in war has been a subject of much discussion. Some authorities have definitely stated that war has no principles; others, when propounding the art of war, have made free use of the word without even understanding its meaning; and still others, those who may be classed as educated soldiers, have made various attempts to establish principles on general inferences, and, as far as I am aware, without much scientific proof.

The necessity and utility of principles is hinted at by Clausewitz when he explains how difficult it is for men excited in battle " to preserve equilibrium of the mind." [1] Yet he does not directly state that the value of principles lies in their power to eliminate self when judgments have to be formed, and so assist us to maintain that mental equilibrium which is only possible when the mind is attuned to the law of economy of force. It is of some interest, I think, to trace this search after principles in modern times.

Lloyd, virtually, lays down three—namely, strength, agility, and universality—which I have already examined. Jackson lays down four. He writes: " The principal points which relate to the management of a military action appear to be comprehended under the following heads. (1) A precise knowledge of what is to be done. . . . (2) A rapid and skilful occupation of such points, or positions, as give the best chance of commanding the objects. . . . (3) The employment of mechanical powers . . . with just direction, united force, and persevering effect. (4) A

[1] *On War*, vol. i., p. 59.

retreat from the contest, when the end is unattainable, in a deliberate and correct manner." [1] Broadly speaking, these may be called the principles of the object, of mobility, of concentration, of offensive power, and of security. Jomini lays down two. He says: "... employment of the forces should be regulated by two fundamental principles: the first being *to obtain by free and rapid movements the advantage of bringing the mass of the troops against fractions of the enemy*; the second, *to strike in the most decisive direction.*" Napoleon lays down no definite principles, yet he apparently worked by well-defined ones, for he once said in the hearing of Saint-Cyr: "If one day I can find the time, I will write a book in which I will describe the principles of war in so precise a manner that they will be at the disposal of all soldiers, so that war can be learnt as easily as science." [2] Clausewitz lays down four: (1) "To employ *all* the forces which we can make available with the *utmost* energy. . . . (2) To concentrate our forces as much as it is possible at the point where the decisive blows are to be struck. . . ." (3) To lose no time, and to surprise the enemy; and (4) "To follow up the success we gain with the utmost energy." [3] Finally, Foch lays down four: "The principles of economy of forces; the principle of freedom of action; the principle of free disposal of forces; the principle of security, etc." [4]

I do not intend to examine these various principles. Some, as it will be seen later on, I consider to be correct, and others incorrect. To examine them would be to digress, since my object in this chapter is to attempt to show systematically how principles are, or may be, derived from the law of economy of force.

If man were so fashioned that he could know all things, he would be omniscient, and if to do all things, then, omnipotent; and, possessing these two powers, he would see that every change which takes place in Nature is righteous, that is to say that it could not in the circumstances take place in any other manner —better or worse.

Man is, however, ignorant, fearful, and weak; consequently, if his aim is to progress, he must seek knowledge, courage, and strength, and the nearer he attains to the fullness of these conditions the more readily will he be able to economize the forces they include. When he has learnt to economize his knowledge, or rather its expenditure, he has discovered wisdom; and when he has learnt how to economize the power of courage he

[1] *A Systematic View*, etc., pp. 23-4.
[2] *Memoires*, etc., Maréchal Gouvion Saint-Cyr, iv., 149-50.
[3] *On War*, vol. iii., pp. 210, 211.
[4] *The Principles of War*, p. 8. What "etc." represents is not mentioned.

has attained to self-command ; and when he has learnt how best to use his strength he has become skilful. The government of these three states is the province of the principles of war.

### 2. The Elemental Base

If the principles of war are to be derived from the law of economy of force, then, as this law controls the changes which take place in the forces of war as expressed by the elements of war when influenced by the conditions which surround them, these principles must be related to the elements themselves. I will, therefore, turn back to these elements and arrange them in what I believe to be the order in which they work.

It will be remembered that I have divided each of the spheres of force into three elements. Thus:

(i.) The mental sphere consists of reason, imagination, and will.
(ii.) The moral sphere of fear, *moral*, and courage.
(iii.) And the physical sphere of offensive, protective, and mobile power.

In each case the third element is the resultant of co-operation between the first two, and also the point of contact with the sphere below it. Thus, force acting on the intelligence causes it to react according to the quality of reason and imagination, and the resultant is will, or the lack of will. Will acting on the sentiments causes them to react to fear and *moral*, and the resultant is courage, or the lack of courage. Courage acting on physical energy causes it to react to pressure (offensive power) and resistance (protective power), and the resultant is movement, or the lack of movement, which takes place in the material sphere outside man. Taking one man as an instrument expressing all these forces, they can be plotted out as shown in diagram 15.

Reason and imagination, in close co-operation, decide on the object and the force to be expended in its attainment. This decision is expressed by the will. The will now enters the moral sphere, and, if *moral* repels fear, the will impinges on courage, and from a purely mental force becomes a moral one. Courage, vitalized by will, impinges on physical energy, which, if the offensive and protective powers are in close co-operation, results in movement.

Man's object is correct action, or action which may be designated as true and not false, therefore truth may be accepted as the governing condition. The nearer action coincides with the true state of things the more correct it will be.

So far the relationship of the elements within man; now as regards their relationship between a general and his troops.

The general is pre-eminently the brain of his army; his main duty is mental, and not physical. With his men it is the reverse, for, though they must use their brains as individuals, as a mass of individuals they, in the main, must make use of their physical

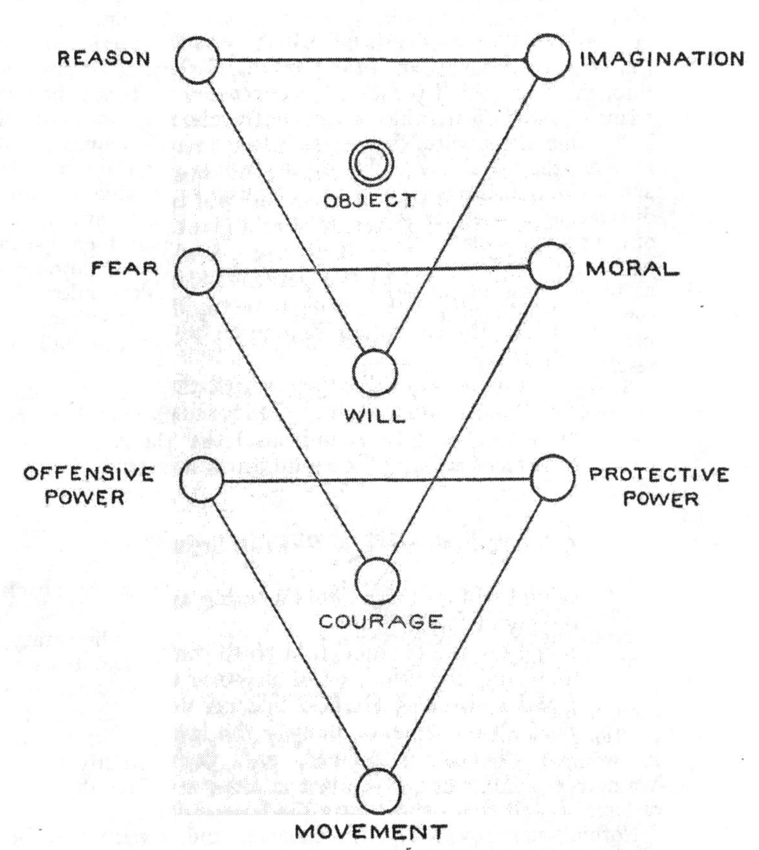


DIAGRAM 15.—THE GENERAL RELATIONSHIP OF THE ELEMENTS OF WAR

powers as directed by the will of the general. As regards moral force, it influences both the general and his men alike, but whilst with the first courage must be active, with the second, in order to accept the will of the general, it must be passive. We can now plot out another diagram (No. 16) which is worth examining.

The general is represented by the upper triangle, and his men by the lower, and these triangles are connected by a line, or bar, which represents the moral sphere of force. We then see that the general must be possessed of a courageous will, a will which expresses self-assertion, the assertion of his plan, which his reason and his imagination have enabled him to formulate; and that his men must be imbued with a self-sacrificing will to move in accordance with this plan, which is rendered possible through their protective and offensive powers. Between the two triangles stands fear, which is the common enemy and ally of both. For, if the will of the general is to control the movement of his men, the moral line, or bar, must, so to say, remain straight and rigid. If thrown out of adjustment by hostile pressure directed against either end, the opposite end will be swung out of the perpendicular. If fear be regarded as a pivot, then if such hostile force is directed against protective or offensive power, so as to push the moral line out of the perpendicular, unless the courage of the general is sufficiently strong to rectify this diversion, moral contact between the two triangles may be broken, and the result is demoralization.

There are many further considerations which these two diagrams suggest, but these I must leave to the student to discover, as my intention here is not to examine all the relationships between the elements of war, but to establish a scaffolding for its principles.

### 3. The Principles of the Mental Sphere

For the time being I will set this scaffolding aside, and turn back to the law of economy of force.

From the seemingly opposite, though in truth complimentary, forces of attraction and repulsion, or of pressure and tension, or of opposition and resistance, Herbert Spencer deduces three laws of direction of all movements, namely the law of greatest traction, the law of least resistance, and their resultant. According to this philosopher, " the last is alone strictly true." We may, I think, call these three laws the laws of the direction of force.

Nothing can move without a direction, and, given force, the whole problem of its economical expenditure centres on the

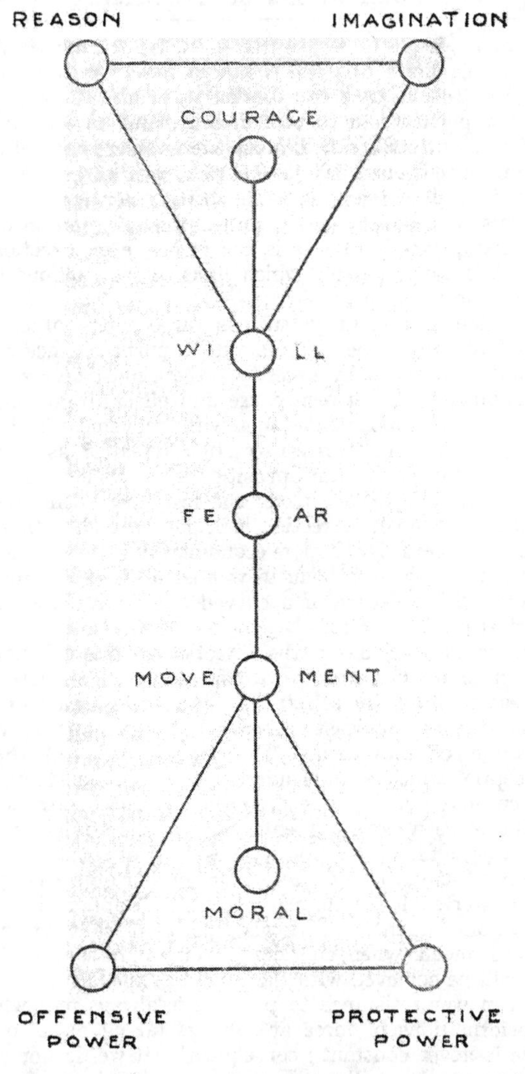


DIAGRAM 16.—THE DUAL RELATIONSHIP OF THE ELEMENTS OF WAR

direction given to it. In Nature this direction is determined by cosmic laws. In war we have nothing so omnipotent or disinterested to guide expenditure of force; nevertheless, we must apply this law in part, if not in full; we must master it, so far as our intelligence can master it, or else it will master us. As our intelligence has to direct force, and as our intelligence is limited, I will not call the abstract conception of direction which should guide us a law, but a principle, and, in my opinion, the first principle of war is the principle of direction of force, and it is this principle which links all our actions to the law of economy of force.

In affairs between men, such as war, economy of force demands that force should be directed with a purpose, since rationally it cannot be directed by necessity—there must be a reason for its expenditure. If between cause and effect (beginning and end) no apparent opposition is met with, movement will take place along the line of greatest traction (greatest assistance) or of least resistance. In war opposition is always met with; therefore movement takes place along the resultant of all tractions and resistances, and its direction is seldom straight—that is, direct. The straighter it is the more economically shall we reach our goal, consequently our problem in war is to direct our force along a straight line in place of a curved one, or a spiral, and in the shortest possible time.

As motion occurs in the direction of the greatest traction, the more we can concentrate force along this line, either by increasing it or by selecting a line along which opposition is weak (the line of least resistance), the less curved will be the direction of movement. Therefore I will call the abstract conception which should guide us in all endeavours to straighten out the curve of our direction the principle of concentration of force.

The more we can concentrate force the straighter will be its direction, and, as this presupposes lack of resistance, the longer will our force last, and the sooner will our object be gained, and the nearer shall we approach to the full application of the law of economy of force. This is what, philosophically, I think, Herbert Spencer means when, considering volition, he says: "The desired end will be achieved with the smallest expenditure of force."

As in war resistance to pressure is always met with, and as transformations of force are always taking place, our tactical force is never constant; consequently, if we do not understand its changes, we shall not be able to rearrange our forces so that concentration is maintained. We may, as it is our intelligence and not cosmic law which is in control, side-track it, or thrust it along a line of resistance, or let it dissipate itself. Therefore, as each transformation takes place, we must so thoroughly understand the cause of it and the value of effects that we can economically redistribute our force; consequently I will call the abstract conception of adapting concentration to circumstances the principle of distribution of force. This principle governs the development of force in war, the integration and disintegration of force; it is, in fact, the military counterpart of the law of evolution, and its compliment the law of dissolution.

If right through a war we know how to distribute our force, unless we are very inferior in force, we shall be able to concentrate superiority of force; and if we concentrate superiority of force we shall be able to direct our force along the resultant of the lines of greatest traction and of least resistance, and, if we can so direct our force, then will our expenditure of force be economical and the law of economy of force will be maintained.

These three principles—the direction, concentration, and distribution of force—are not only co-equals but inseparable instruments of the mental sphere, and through the mental sphere of the moral and physical spheres. They can be infringed individually or collectively, but they cannot be annulled, for they govern the machinery of the engine of war, the output of which is economy in varying degrees. Though these modes of the law of economy of force—for such they in fact are—must be set in motion by the will of man, the hand which holds the throttle of this engine is cosmic law, which operates without let or hindrance, irrespective of man's wisdom or folly.

To turn back to the elements. Whilst the interplay between the ideas is imagination, and whilst imagination is ceaselessly shuffling ideas to and fro and weaving them into all manner of designs, according to the object which is at the moment in control of the mind, reason is simultaneously selecting such of these designs which, when fitted together, like the pieces of a puzzle, will make a complete picture of our intention. Once this picture is completed the will is released. The picture now may be compared to a map, the will to a man, and the action resulting to finding his way from place to place across country by means of this map. The shortest way from place to place is in a straight line (a curve on a globe). Does the map correspond with geography (reality)? Has the imagination grasped what the surface to be traversed is like? Has the reason worked out the shortest, that is, in the sense I make use of this word, the most economical road; and is the will strong enough to travel by it? These are the questions we must answer if our aim is the correct application of these three principles, and the last of these answers brings us to the moral sphere of war.

### 4. The Principles of the Moral Sphere

The mental endeavours of the general and of each of his men, when engaged in individual action, are concentrated on the discovery of the most economical line of direction. The initial impulse is the object, and the magnet which attracts the will is the objective, and the vibrations between these two poles must, if they are to be economized, travel by the most direct line; this presupposes action.

This action, which must eventually be developed in the physical sphere, will be resisted by the enemy's physical force, and must consequently be opposed—that is, pressed back—by a similar force, which depends for its endurance on the strength of the moral sphere separating the mental and physical spheres. The direction taken by the will must, therefore, traverse the moral sphere before it can set in motion the physical.

If men are controlled by fear, they will not move, or, if they do, their movements are likely to be chaotic. The more courageous they are the more directly will the will of the general be able to control their actions. This condition of courage depends, as I have shown, on how far the resistance of *moral* can keep at arm's length the pressure of fear; therefore the conditions in which direction is asserting its influence must permit of the development and maintenance of the maximum active courage from the initial or potential courage of the army. The degree of this courage, consequently, determines the quality of the action resulting, therefore I will call the abstract conception of the potentizing of the will of the general by means of his courage and that of his men the principle of determination of force.

The strength of the moral sphere of force is, as we see, largely dependent on the correctness of the line of direction decided on by the will of the general, or man acting individually, consequently on the principle of direction of force depends the moral pressure of the instrument. Its tension, or resistance, depends on its initial moral value, the training it has undergone previous to action. Hostile resistance attempts to frustrate its pressure, and hostile pressure aims at overthrowing its resistance.

In the mental sphere I have shown that direction of force is dependent on concentration and distribution of force; consequently, if harmony is to be maintained throughout the entire forces of the instrument, concentration and distribution of force must equally be applied in the moral sphere.

Moral pressure depends for concentration on the line of direction taken; therefore the question which must be answered is,

" What should be the aim of this direction ? " The answer is that the aim should be the breaking down of the determination of the enemy's command or instrument, by so demoralizing it that its *moral* is unbalanced by its fear, and the union of the elements of will and courage is broken.

In an expected attack the resistance to be met with will obviously be greater than in an unexpected one, and the less the resistance the greater comparatively will become any given amount of pressure directed against it. Consequently in the moral sphere concentration of force is represented by surprise, therefore I will call the abstract conception of moral concentration of force the principle of surprise, or the principle of the demoralization of force.[1]

Distribution of force in the mental sphere must also have its counterpart in the moral sphere. The moral resistance of the instrument must frustrate or withstand the moral pressure exerted against it and resulting from the enemy's physical action. What will be the direction of this pressure—that is to say, what will be its line of approach towards overthrowing its adversary's determination ? We cannot say. But if it is to our advantage to surprise the enemy, it is equally to his advantage to surprise us. We cannot distribute our *moral*, for *moral* is not a commodity, but we can so distribute our men that an unexpected attack will be unlikely, or most difficult ; further, we can distribute them in such an order that no single party is isolated, and, consequently, lacks, if not immediately the physical, then the moral support of the whole or of other parts. Again, we can, by training and education, distribute a high moral throughout our force, and so endow it with power of enduring the pressure of both expected and unexpected hostile action. Consequently I will call this abstract conception of the distribution of moral force the principle of endurance of force.[2] On the ability to apply this principle, and simultaneously bring into operation the principle of surprise, will depend the economy of our determination. Hence, as direction of force depends on concentration and distribution of force, so does determination of force depend on demoralization (surprise) and endurance of *moral*.

Again, these three principles are not only co-equals, but

[1] As surprise so frequently is accomplished by an unexpected move, originality of thought and novelty of action are potent modes of this principle.

[2] As originality and novelty play an important part in the application of the principle of surprise, so do simplicity and common doctrine play an equally important part in the application of the principle of endurance. An original plan should aim at simplicity, and novel action should not demand movements the troops do not understand. If these four requirements—originality, novelty, simplicity, and common doctrine—can be closely combined, then determination will be strong, but, if not, it is liable to prove fragile.

inseparable instruments of the moral sphere, linking as they do the mental to the physical sphere, and they constitute the moral modes of the law of economy of force.

In these principles (just as in the mental ones) we see the interplay of the elements of the moral sphere. Direction having laid down our road, the progress along it depends on our encouraged will; fear springs up everywhere, for it is, in fact, the atmosphere of the battlefield, a poisonous gas which, if we breathe it, will asphyxiate our courage. To take a simile, our gas-mask is our *moral*, and as long as it remains in an efficient condition, so long will our courage endure; but should it prove defective, or should the enemy's action injure or destroy it, then courage will slacken or die, and the contact between the will of the commander and the actions of his men will be broken.

### 5. The Principles of the Physical Sphere

I now come to the physical sphere, the sphere of true action. The encouraged will, expressed by the principle of determination of force, must set the military instrument in movement, whether this instrument be one man controlled by his own will or an army controlled by the will of its general. Movement depends on physical energy, and how far this energy is concentrated or dispersed. If the direction towards the objective is simple, then through physical energy can force be concentrated against it; if complex, then force must be distributed, and the various movements resulting must be correlated. The degree of movement, consequently, directly depends on the pressure exerted and the hostile resistance opposed to it, and also on the determination shown, which depends on the moral endurance of both sides, and the freedom of this endurance from surprise. Finally, movement must coincide with the direction decided on, for movements away from this direction are eccentric to the plan, and are, consequently, destructive to the will of the general. Movement must, therefore, express the will of the general through the will of his men, their determination acting on their physical energy; the abstract conception of such movement I will call the principle of the motion of force, or of mobile action, or simply of mobility.

As movement in war is met by resistance, it must be expended in the form of pressure. This resistance depends on the determination of the enemy; but this determination is itself dependent on the physical organization in which it is encarded. This organization possesses structure, maintenance, and control, all of which are organically essential ingredients. Pressure can be exerted against any one of these, or, more generally, all three simultaneously. Normally the process whereby pressure is exerted is to concentrate a superiority of physical force against the structure of the enemy's army, and attempt to destroy or disorganize it. The abstract idea of such action I will call the principle of the disorganization of force, or of destructive action, or of the offensive.

If pressure is exerted against the body of the enemy's army, destruction of force becomes direct, and this has been the normal method throughout the history of war; if against his system of maintenance, it becomes indirect; and so also if it is directed against his moral endurance, or the will of his general. In the first case pressure manifests in fighting, the object being physical destruction; in the second it takes the form of physical disorganization through economic pressure; in the third, of demoralization through surprise or terror; and in the fourth, to a similar end through similar means, but directed against the will of the general rather than against the will of his men.

As all these forms of pressure can be exerted, it stands to reason that to concentrate physical force alone is not sufficient. However carefully a plan may have been worked out, however thorough has been the reason, however illuminating the imagination and decisive the will, no general is omniscient and no soldier omnipotent, consequently the possibility of error in direction always exists. Therefore, besides concentrating our physical force, we must also distribute it in such an order that structure, maintenance, and control may be maintained. The major tactical distribution must be such that, through a combination of formations, the economy of the plan is maintained, and the minor tactical distribution must aim at protecting pressure whilst it is being exerted. I will call, therefore, this abstract idea of physical endurance of force the principle of security of force, or of protective action, or simply of security.

The more pressure is secured by resistance the greater will be the mobility, or potential motion, of the instrument; thus mobility is dependent on the co-operation of these two, and it is the effect produced by this co-operation which is its cause. Economy of movement—that is, doing something in the shortest time, with the least loss of energy, mental, moral, and physical—is the ultimate expression in battle of expenditure of force. If movement were absolutely perfect, it would coincide with the law of economy of force. Thus the final principle of war—mobility, which is the resultant of the co-operation of the previous eight, working as parts of an engine—is the link which unites the final effect with the originating cause, and the closer the coincidence between these two the more perfectly has the law of economy of force been applied. Diagrammatically this may be shown as in diagram 17.

We start with an object, which presupposes an objective. Our directing law is economy of force, our means are our instrument, which is governed by the nine principles of war, which are, so to speak, emanations of the one law as applied by our intelligence.

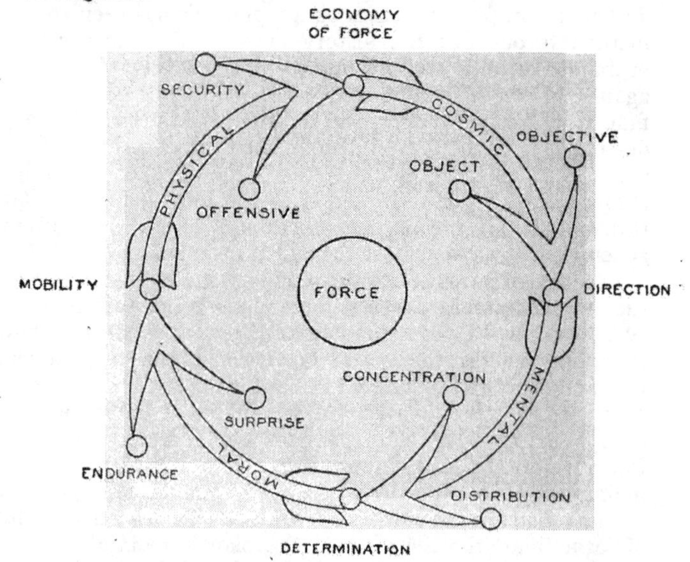


DIAGRAM 17.—THE UNITY OF THE PRINCIPLES OF WAR

We see in the principles of the physical sphere the interplay of the elements of this sphere. As *moral* was our gas-mask, weapons are our offensive tools which overcome resistance, and, as they clear the road, we move along it, and protection is the glove which covers our hand.

## 6. The Principles of War

In the preface of this book I outlined the history of the principles of war as reasoned out by me, and there I examined the differences between my earlier and present conceptions. I do not here want to repeat these differences, but, as I have in the present chapter given more than one name to several of the principles, I think that it will be as well if I now decide on one name for each.

As they are emanations of the law of economy of force, in my opinion the following are the terms which more scientifically express the energies they control:

(i.) The principle of direction of force.
(ii.) The principle of concentration of force.
(iii.) The principle of distribution of force.
(iv.) The principle of determination of force.
(v.) The principle of demoralization of force.
(vi.) The principle of endurance of force.
(vii.) The principle of mobility of force.
(viii.) The principle of disorganization of force.
(ix.) The principle of security of force.

These terms have, however, certain disadvantages, the main one being that in our army other names are being used for several of them; I think, therefore, that the most practical, if not the most expressive, terms are:

(i.) The principle of direction.
(ii.) The principle of concentration.
(iii.) The principle of distribution.
(iv.) The principle of determination.
(v.) The principle of surprise.
(vi.) The principle of endurance.
(vii.) The principle of mobility.
(viii.) The principle of offensive action.
(ix.) The principle of security.

And as such I will usually refer to them.

I will now arrange these principles in two diagrams, in the manner I adopted for the elements of war. In the first diagram (No. 18) I will show the principles working within man, and in the second (No. 19), between the general and his army.

In the case of one man the problem, in brief, is to discover the relationship between the object in his mind and the objective which confronts him. For example, a man wishes to pick an apple; the obtaining of the apple is his object, and the apple itself the objective. In war the political object is a better peace, but the military object is to establish a condition which will permit of this better peace being attained; the objective is the

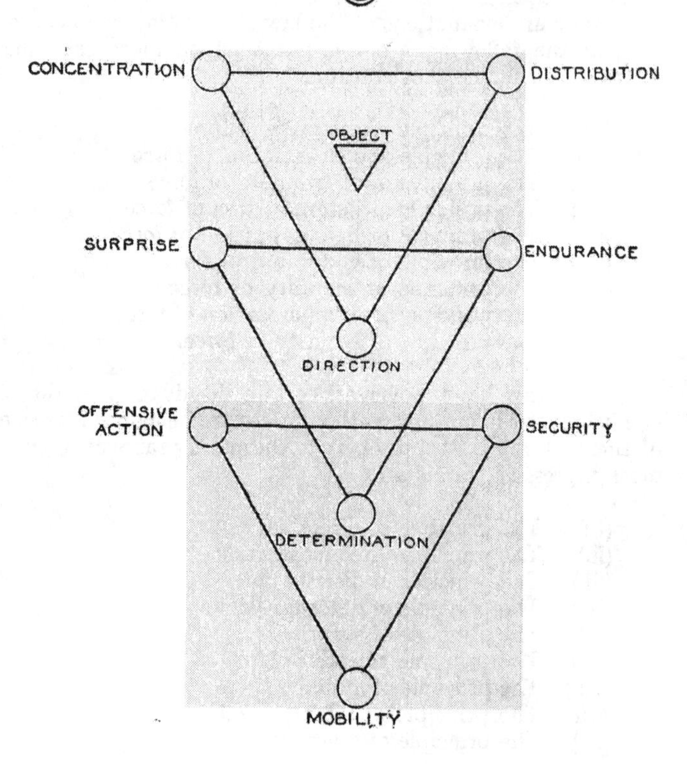

DIAGRAM 18.—THE GENERAL RELATIONSHIP OF THE PRINCIPLES OF WAR

disarmament of the enemy, which demands the occupation of his country.

To return to the man and the apple. The apple, I will suppose, is on a branch out of reach. There must be one way out of all ways in which the least amount of physical energy need be expended in obtaining possession of it. Which way is this? By distributing mental force—that is, by using our imagination—we shall see that there are several ways of climbing the tree, and, guided by the idea that we should economize our force, we select as our working hypothesis one way. We hand this over to the reason, which analyses it, and, after having concentrated thought on the idea, accepts it, rejects it, or amends it. Finally, between imagination and reason is built up a synthesis, or a plan of action, the completion of which releases the will, which gives it a definite operational direction.

The man now approaches the tree; to climb it will not demand more thought, but determination, in the present instance a will to climb, which means that his pluck must cancel out his fears.

He starts to climb the tree, which means that he must secure himself, perhaps with both hands to begin with, by grasping the branches; but eventually his security must be such as to leave him one hand free to seize the apple—and I will suppose his left hand. His movement depends, in fact, on his security.

He stretches out his hand to pluck the apple, but he has not noticed that a wasp has settled on it. This insect stings him. Surprised, fear is awakened, which in an instant has cancelled his pluck (moral endurance). His determination vanishes, and, with his determination, his will to seize the apple, and, with loss of direction, his reason and imagination are momentarily blotted out. He jerks his left arm backwards, which causes him to wrench at the branch he is holding on to with his right hand. The branch snaps, and he falls to the ground.

Now as to the second diagram (No. 19), which depicts the principles working between a general and his army.

A farmer wants to obtain an apple which, again, is on one of the top branches. After looking at the tree, he calls to him a boy and tells him how to climb up it. Though this order relieves the boy of making any extensive use of his brains, he has to use them to a certain extent. The boy begins to climb the tree, but soon gets into difficulties, and shouts down that he cannot climb any higher—in fact, his pluck is giving out. The farmer is, however, determined that the apple is going to be his, so he shouts back: " If you do not get that apple I will thrash you." This stimulates the boy to climb higher—an offer of twopence might have done likewise, or even an encouraging word. As the boy

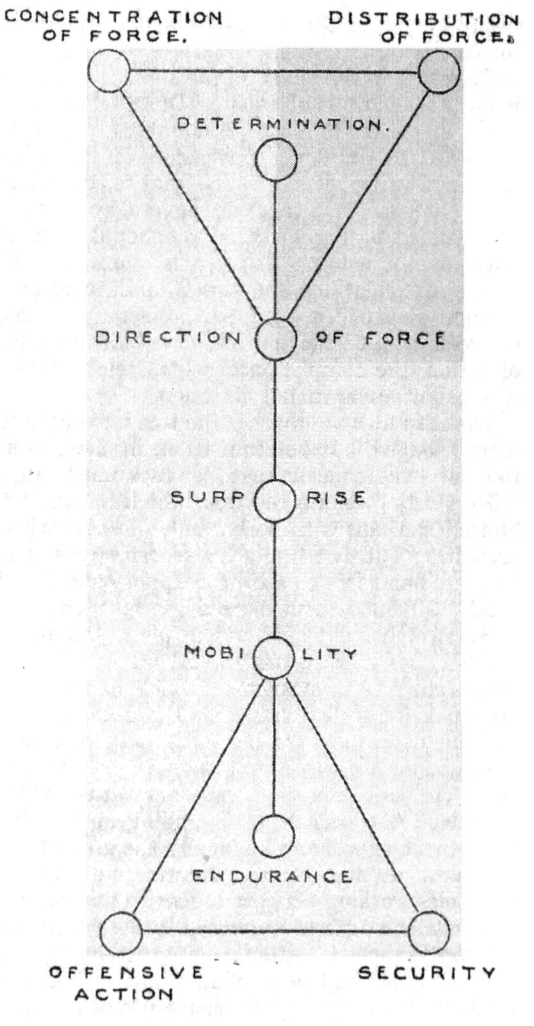


DIAGRAM 19.—THE DUAL RELATIONSHIP OF THE PRINCIPLES OF WAR

nears the apple the farmer, who has been watching him, cries out : " Take care, there is a wasp there. Here is a stick. Knock the apple off." The apple falls to the ground.

I have taken these two very simple examples because they have nothing directly to do with war, yet indirectly they have *everything* to do with it ; for, whether we are attempting to gather apples or kill men, *the principles which govern our actions are the same.* I need not, therefore, elaborate these examples into military operations.

The first thing to remember is that, whether we are working on our own or with others, these nine principles, if correctly applied, assist us in attaining economy of force. The second is that, in spite of this, when we are working directly under another to a large extent we are relieved of mental work ; the plan is given us, and, though we must use our intelligence in carrying it out, our direction in doing so is strictly limited. To the general the principles of direction, concentration, distribution, and determination are all-important, and to his men so are the principles of mobility, offensive action, security, and endurance. Surprise is common to both.

There are many ways in which these principles can be arranged, and they depend on the individuality of the student. Ultimately they are all of equal value, since all nine are essential to economy of force. The simplest method of employing them is, I think, one which ranges them in three groups, under control, pressure, and resistance. Thus economy of force controls direction through the pressure and resistance expressed by concentration and distribution ; in its turn, direction controls determination through the pressure and resistance of surprise and endurance ; and determination controls mobility through the pressure and resistance of offensive action and security. The three groups are therefore :

(i.) *Principles of Control :* Direction, determination, and mobility.

(ii.) *Principles of Pressure :* Concentration, surprise, and offensive action.

(iii.) *Principles of Resistance :* Distribution, endurance, and security.

We thus obtain a threefold order of control springing from a dual order of pressure and resistance, each of these dual forces being in itself a threefold one. Ultimately these three groups form one group—economy of force.

In the following three chapters I shall examine these three groups.

### 7. The Constructive Value of Principles

A little while back I stated that the principle of mobility was the resultant of the co-operation of the remaining eight; co-operation, therefore, in the full meaning of the word, is the tangible expression of economy of force. If co-operation could be perfect, it would mean that we should be able to obtain the fullest possible economy; it cannot, however, be perfect; nevertheless the nearer it approaches perfection the more correct will our actions be. By examining a battle, a plan, or an organization we can discover how far actions, ideas, or parts were, or are, suitable to co-operate, and by this process of analysis we can discover facts of importance—the reasons for the errors made and for the successes gained.

Co-operation always exists, but in what degree? This is the problem. Is it as effective and economical as it can be? How can we discover its value? The answer is by examining its ingredients, namely the elements of war. Are the principles animating them in harmony, or is any one discordant? Why is it discordant, and how can its faults be remedied? By such questions we arrive at scientific answers.

In order that the student may fully understand the inner working of co-operation, I will take a simple concrete example—a clock. If we open up its works and look at them, we shall see a small spring vibrating, and a larger one apparently motionless, but, in fact, slowly unwinding itself; certain wheels move rapidly, and others slowly. There are three main parts—a mainspring which releases concentrated power, a hairspring which controls the output of this power, and a system of gears which distributes this power. We obtain, in fact, a close similarity to concentration, direction, and distribution of force. The whole mechanism is working in unison in order to move the hands of the clock over the dial, so that an observer may read the time of day. A good chronometer will not lose or gain more than a few seconds a year; its economy of force is almost perfect; yet it has been made by man to assist man. So with an army, though we cannot construct a military instrument as economically as we can a watch, we can at least attempt to set its parts together in such an order that a fair degree of unity of action will result.

Such is co-operation, or working together for a common object, in the case of the watch for the registration of time; yet it is only co-operation in a sense, a general sense, for, suppose that the dial had no figures marked on it, the most perfect co-operation would prove useless. And suppose again that, though the figures are there, the clock gains ten minutes in every hour, the reading of time will be misleading, and the objective of the reader will not be gained.

I will translate this into military terms.

The correct registration of time is the object, its correct reading the objective, and the dial is the plan. The works are the elements of war whereby the object is gained, and as these works are governed by mechanical principles based on pressure and resistance and their resultant, so are these elements controlled by the principles of war which possess a similar foundation.

### 8. The Development of a Plan

I will now turn to planning, and examine how an economical plan is worked out.

Means must be scientifically fitted to ends according to conditions; the foundation of every plan is, therefore, common-sense action. Thus in land warfare our means are our army, national resources, national *moral*, etc.; our end is the enforcement of our policy by defeating the enemy at the least cost to ourselves, not only in men, but in resources and honour; and the conditions are the innumerable factors which are met with in war, many of which are continually changing.

How are we going to control these conditions? Here, then, is our main difficulty when we attempt to devise a co-operative plan—that is, a correct placing of the figures on the dial so that the energy generated by the mechanism of our military clock is economically expended. By plan I do not only mean large arrangements such as those suitable for an army, but equally smaller ones down to those suitable for a platoon, a section, or even one man.

To control conditions—and on their control depends the structure and maintenance of the instrument—we must work as follows:

First, we collect all the conditions we possibly can, and arrange them according to their elemental categories—this is information. Secondly, we apply to these conditions the principles of war—this is analysis, which will enable us to discover which conditions will assist us and resist us. This leads to transformation, and through transformation to hypothesis. Thirdly, having ascertained the military values of the conditions, bearing the hypothesis in mind, we equate these values with the elements of war and discover how these elements will be affected. Then, bearing in mind our object, guided by the principles of war, we arrange these elements and set them together in a plan—this is synthesis.

Every individual worker, according to the particular " mould " of his brain, will work somewhat differently, for working is an art, but the foundation of art is science. Hypothesis, analysis, and synthesis may be compared to a triangular frame which holds the " substance " of our thoughts, and the ego itself is the worker.

I will now take as an example the simple platoon operation I examined in chapter viii. A platoon is ordered to capture a machine-gun post. This is a very simple operation, yet it may be a very dangerous one ; in any case, the platoon commander is directly influenced by immediate danger, and, unless he can maintain complete self-command, he is apt to base his advance on an unsound plan, or no plan at all. He has three main things to think about—the ground, his men, and their weapons from his point of view, and from that of the enemy.

He first wants to discover the most economical direction of advance, so he rapidly examines the ground. How will it influence his men and their weapons? He must start from a secure base ; but the heavier the blow he can deliver the more rapidly will the objective be won, so he thinks in terms of concentration of force—the knock-out blow. But he must not put all his eggs into one basket, for he must never forget that not only must he distribute part of his platoon to protect his decisive attack, but that, however carefully he may have analysed conditions, new ones will always be cropping up, to meet which he must hold a reserve in hand ; consequently he must apply the principle of distribution of force, which not only means holding men in reserve, but also forcing the enemy to disperse his men by threatening him from more than one direction. By now he should have a fair idea of his direction.

Having settled this point provisionally, he secondly considers his men and the probable condition of the enemy's. Are they tired or fresh, are their tails well up or down ? For ultimately it is the men who have to take the position. Is their any chance of delivering a surprise blow, perhaps by pushing a couple of riflemen round a flank ? The ground will help to answer this question, but still more so will the determination of his men ; what is their endurance ? He is now thinking in the terms of the moral principles of war, and they will enable him to check his provisional direction, and perhaps improve it.

He has now determined on a course of action, so he turns to the physical principles. He knows the condition of his men and the state of the ground ; how now can he move over it ? The principle of security applied to its protective characteristics in relation to the protective power of his weapons will tell him where he can best resist the enemy, and the principle of offensive action applied to the offensive characteristics and powers of ground and weapons, where he can best exert physical pressure against him. These two combined will, when equated, enable him to decide how to move. His direction is now fixed, and the action begins.

I do not pretend that any platoon commander will normally have time to consider the principles of war so methodically. I have made one principle follow another in a logically stereotyped order. But, if he has trained his mind to think in principles, in place of thinking by order of conditions, directly he thinks of one principle he will think of the influences of the remaining eight. As conditions change, he applies them, and the quicker he can do so the higher will be his initiative, and by initiative I do not mean doing something, but doing the right thing—the common-sense thing. Thus is economy of force observed, and each small economy effected adds to the ultimate victory, or minimizes the ultimate defeat.

# CHAPTER XII

## THE PRINCIPLES OF CONTROL

Life is not a bully who swaggers out into the open universe, upsetting the laws of energy in all directions, but rather a consummate strategist, who, sitting in his secret chamber over the wires, directs the movements of a great army.—BALFOUR STEWART.

### 1. THE PRINCIPLE OF DIRECTION

GRANTED that the general is a free agent, and is not cramped in his action by political pressure, and granted that he is a man of normal intelligence, how should he proceed if his aim is to establish a condition which will result in the gaining of the political object of the war?

One writer says: "The ordinary Englishman places too much confidence in imperfectly directed energy, vulgarly called 'British pluck,' and too little upon the fundamental knowledge which should direct it." This fundamental knowledge demands an understanding of the forces commanded and of the conditions in which they will be expended. This writer further says: "Hence, when men wish to effect objects, they must first adapt themselves to the energies and conditions which govern them in the particular case";[1] and this they cannot do without knowledge, knowledge gathered from history and study before the outbreak of the war, and by the intelligence and reconnaissance services during it. Further, they cannot do this unless they know how to adapt themselves to the various changes of a campaign or battle. The first requirement is, therefore, knowledge; the second, the understanding of the items of knowledge and their relationships, and the third is wisdom in the application of this understanding, and here it is that the principles of war come to our assistance. To know, to understand, and to apply wisely are the three closely related means of arriving at a plan of action, and they are none other than our old friends observation, reflection, and decision, in more general forms.

"What have I got to do?" The answer to this question is the starting-point of every plan. "How am I going to do it?"

[1] *The Scientific Basis of Morality*, G. Gore, pp. 123, 118.

And the answer to this one leads to the line of direction of every plan. If a platoon commander is ordered to capture a hostile machine-gun he knows what he has got to do, and he can arrive at the answer to the second question by estimating his and the enemy's forces, and by relating them to the advantages and disadvantages of ground, positions, and time at his disposal, etc. The larger the forces to be employed the more difficult grows this problem, until, when we arrive at the general-in-chief, the problem becomes one of immense complexity, normally rendered worse confounded by the fact that the politician does not tell him the nature of the political object of the war. Is it to annihilate the enemy; is it to break his will and spare his industries; or is it to result in a condition which will engender no vindictiveness after war is concluded? What is it? What does the politician want? And, again, what is the maximum price the nation is willing to pay for the gaining of this object? Will the nation "pawn its last shirt" in order to win the war, or how far will it go? If the war is to be fought to a finish, regardless of cost, is the Government immediately prepared to mobilize the entire resources of the nation, and, if not, then how much, and when?

If a general is informed on all these points and many others he will know how to base his plan on policy; he will know, not only what he has immediately got to do, but what he eventually will have to do, and he will be able to direct his forces, not only according to enemy pressure, but in relationship to their future development.

Given this information, a general can base his plan on policy; but, if not given it, he must act on his own, and hope, against belief, that he will not be interfered with politically.

## 2. The Line of Direction

In his plan he should aim at establishing such a condition that policy can take effect. Normally this condition demands the annihilation of the enemy's resistance and the occupation of his country. These conditions can only be secured by strategical, tactical, and administrative action. The strategical object is to gain freedom of movement; the tactical, freedom of action; and the administrative, freedom of supply. The first is gained by correct distribution; the second by superior concentration, physical and moral; and the third by secure communications. These three combined will give him his direction, and they can never be separated.

Whatever the circumstances may be, our action depends on the enemy's action, which, in its turn, depends on our action. Military thought can, therefore, seldom, if ever, be directed in what may be called a straight line—that is, without interference from starting-point to goal. In 1870 von Moltke based his plan on " general direction, Paris; objective, enemy wherever met." This, I think, in the circumstances, was sound strategy, for not only was Paris the political centre of France, but by directing his forces on to Paris he compelled his enemy to interpose, and so forced the French to battle, and he wanted battle. Had, however, the French been able to concentrate north and south of the line of his direction, namely Metz—Paris, he would have had to change his direction, and, unless his plan admitted of this change, it might have broken down. In 1914 von Moltke the younger, following, at least in part, the plan of von Schlieffen, directed his main forces from Liége on Paris, because he hoped on this flank to avoid a frontal battle with the main French forces, his object being to attack them in rear from the direction of Paris. His objective was first the French line of communications from Paris eastwards; and only secondly the French armies in Alsace and Lorraine. In 1494 Charles VIII, because of his preponderating strength in artillery, took a map, and chalked on it the exact places he wished to go to, and he went to them irrespective of the enemy's action; because, if the enemy appeared on his line of advance, he simply blew him off it.[1]

The operations of Charles VIII are the exception to the rule that direction in war is never straight, but in place curved, yet they show that the straighter our direction becomes the simpler is the problem, and this straightness depends almost entirely on pressure.

What a general would like to do would be to exert pressure in one definite direction; but normally the enemy prevents this, not only by resisting pressure, but also by pressing in some other direction; consequently his final direction is the resultant of the general pressure and resistance of his own forces and the enemy's. I will now turn to some of the more important conditions of war which influence direction.

### 3. The Point of Direction

What is the point of main pressure, or the decisive point against which pressure can attain the most economical results? It is not necessarily the line of least military resistance, since the

[1] Machiavelli said: " He conquers Italy with a piece of chalk."

military forces are only part of the enemy's instrument of war. His political control is centred in his capital and his power to maintain his fighting forces in his industrial areas. A Government can change its seat, as the French Government did in 1914, though this is apt to demoralize the nation ; but industries cannot change their localities. For example, it is obviously impossible for the Germans to move the Ruhr coal-fields into Silesia, or for ourselves to move the port of London to Bristol. Consequently, as fighting forces are becoming more and more dependent on industry, industrial areas are becoming great magnetic centres of pressure. Pressure is steadily being attracted towards them, and in the future great battles will undoubtedly be waged to win or to hold them. This will lead to war in thickly populated areas.

Turning now from industrial and political conditions to military ones ; where is the decisive point to be sought ? This again is by no means a simple question, and the simplest method of arriving at an answer is, I think, to examine this problem from the point of view of the three forces of war.

One general wishes to defeat another general, and, until comparatively recent times, as I have already stated, the death, or capture, or serious wounding, of either general normally decided the day ; for the general *was* the plan. He could personally direct his troops, and, according to circumstances, he adapted his thoughts and applied his actions. To-day the general devises, or should most certainly devise, the plan. He is no longer in physical control, and, once his plan is issued, the mental structure of command is enlarged to include a number of subordinates, who, if they are capable men, can, in an emergency, replace him. The mental decisive point is, therefore, the enemy's plan, which holds his decision, and, if this decision can be revoked, mentally the enemy is reduced to a state of reflection—that is, of reasoning in place of willing. He has to reason out new moves before his men can execute them, and, consequently, loses time. Conversely, his antagonist gains time, and, gaining time, can make more use of space and all that space includes, namely the conditions of war. The decisive mental attack is, therefore, directed against the enemy's decision as expressed in his plan. If the enemy's plan is known his decision can be discovered ; it is, however, seldom known, but it is frequently discoverable, if the character of the commander has been previously analysed, and if the national characteristics and tactics of the enemy have been examined, and the geographical conditions of the theatre of war are understood. Alexander grasped quite clearly what Darius was worth, and he defeated him in every battle ; the

Romans could never grasp what Hannibal was worth, and they sustained defeat after defeat. The pivot of the enemy's will is his plan, and, if this plan is smashed, the chances are that the enemy is smashed. Therefore in war the mental line of direction is towards the vital point in the enemy's plan.

Morally, the vital point is the rear of the enemy's army, because the enemy is least prepared to sustain pressure in this direction, and, if pressure is exerted, he will almost certainly be compelled to abandon his plan until he has successfully secured himself against this pressure or destroyed it. Force directed against the plan compels the enemy to reflect; but force directed against the rear of his army compels him to change his determination, and the rear of an army is the morally vital point, because the army is the instrument of his plan.

There are several ways of carrying out this attack:

(i.) By enveloping a flank.
(ii.) By penetrating a front.
(iii.) By manœuvring an enemy into a position which opens his rear to direct attack.

To-day aircraft, if in sufficient strength, can always attack the rear of an enemy's army; and fast-moving tanks and armoured cars will equally well, and even more directly, be able to do so, if the enemy relies on infantry as his main arm. Aircraft can attack, not only the rear of the enemy's army, but the national will this army is protecting, as well as industrial and political centres; consequently, the front of an army no longer protects its rear—or nothing like so fully as it did a few years ago.

Physically, the decisive point is the arm or position which is essential to the execution of the enemy's plan. Thus, if a general determines to occupy a position by means of infantry, led by tanks, the decisive point his opponent should aim at is the position of the tanks. If he intends to occupy a position by means of cavalry in order to sever his enemy's communications, the decisive point is where the cavalry is, at a distance from or on the position itself. In his battle front such positions, the loss of which will compel him to change his plan, are also decisive points, and all these decisive points and actions are dependent on the conditions of war.

A correlation of all these various lines of direction gives the general tactical direction of the plan, and any action which aims at changing this direction is one of a decisive nature. As the will of a general finds expression in the mobility of his men, so does direction ultimately find expression through the principle of mobility. To force an enemy to change his plan, the most general means adopted is to restrict his power of movement, strategical, tactical, or administrative.

### 4. The Organization of Direction

I have in chapter viii. stated that strategical movements are mainly protective in nature, and tactical movements offensive, and in battle I have called these two expressions of movement the approach and the attack. The resultant of these two moods of force is to establish a condition of movement free from all hostile pressure. This condition is administrative movement, and the nearer it approximates to the movements which take place during peace-time the more directly can an army be supplied and controlled, and, consequently, its structure maintained and rejuvenated.

The aim of strategical action is, therefore, not only to direct an army so that the greatest tactical effect is obtainable, but also to direct it in such a manner that its administrative movement, and all this movement includes, is in no way jeopardized, and, if possible, is rendered still more secure. Strategical movement is, therefore, dependent on two important and extensive series of conditions, conditions which affect tactics—fighting—and those which affect maintenance—supplying. In the past, judiciously directed tactical power has normally protected administrative movement, but the introduction of aircraft has seriously modified this protection, since to-day air action can be directed against the rear of an army without serious interference from the ground. As the rear services of an army are as important to it as are the internal organs to the human body, and as in the body these organs are centrally placed between the limbs which are more closely dependent on the ground—the legs—and those which are independent of the ground and above it—the arms—it is more than probable that the future will see aircraft being extensively employed as arms, and not only in advance of an army, but in its rear, the direction of air force units so employed being relegated to the quartermaster-general or his representatives.

I have introduced this seeming digression with a definite purpose, namely to show the complexities which exist and have to be smoothed out before a mean direction can be ascertained which will permit of strategical, tactical, and administrative movements co-operating economically.

The conditions which, in the past, have mainly influenced the strategical application of the principle of direction have been those of communications—roads, railways, rivers, and canals.

Though these conditions are likely to endure, others are rapidly rivalling them in importance.

The object of superior mobility is not only to move more rapidly in a given time, or over a given space, than the enemy, but to obtain a maximum, and, if possible, superior offensive power when the enemy is met with, which as a corollary equally demands that, when the battle takes place, the minimum force of troops is required for protective duties. On the nature of these troops depends their offensive and protective powers. Thus, for example, in the days when cavalry was the superior arm, ground suitable for cavalry action was sought after; when artillery was superior, fortresses assumed a predominating influence; when infantry was superior, battlefields were chosen from the point of view of the musket or rifle. If offensive action was desired, open ground was sought for; if defensive, then enclosed. We thus see that the superior weapon of the day determined the tactical value of conditions, especially physical ones, and that, as strategy has as its object the economical distribution of troops for battle, these conditions largely influenced strategical direction.

In brief, we may say that strategical direction is the resultant of tactical pressure and administrative resistance, and of all the conditions which influence tactics and administration. Equally is tactical direction the resultant of administrative pressure and strategical resistance, and administrative direction of strategical pressure and tactical resistance. Thus, if I want to move an army from A to B in order to engage an enemy in battle at B, then:

(i.) My strategical direction depends on the degree of offensive power I can exert at B, and the degree of protection I will have to allot to my administrative services in getting to B.

(ii.) My tactical direction will depend on the facility of my administration, and the fewness of the men and means I have to allot in order to protect it, and on the security of my force from hostile action during strategical movement.

(iii.) My administrative direction will depend on the power of my strategy to compel the enemy to change his plan, and on the resistance my army can develop when the enemy is met with.

What does this mean? It means that, when we examine conditions, it is not sufficient to extract from them their influence on strategy, or on tactics, or on administration separately, but on all three combined, and that unless this is done we cannot begin to contemplate deciding on our direction. All three should move along one line, but all three want to move along separate lines. What, then, is the mean line, or what I will call the line of harmony? It is the line decided upon by the general which will enable him to develop against his objective superior protected offensive power in the shortest time—not necessarily the maximum power, but sufficient to attain his object; if only because a sportsman who wishes to shoot snipe does not fire half-inch bullets, nor does he rely on No. 8 shot when he is hunting elephants. He does not expend the maximum of force, but a sufficiency of force.

If we examine history we shall find that this line of harmony has seldom been worked out scientifically. I will take as an example the third battle of Ypres. The strategical direction decided on was to advance from the neighbourhood of Ypres towards Bruges and Ghent, the object being to capture or cut off the German submarine bases. The conditions were adverse to tactical pressure, for not only was the ground cut up by hedges, dykes, canals, etc., and covered with farmsteads, but the German right flank rested on the sea and the left on Lille, a large centre of communications. As if this were not sufficiently disadvantageous, the natural resistance the country offered to administrative movement was multiplied a hundred times by destroying the surface of the ground and the drainage system which intersected it by artillery fire. In its turn, tactical direction was limited by a want of administrative pressure, for the supply of the army became, not only difficult, but impossible, and, though we were so placed as to be almost immune from strategical interference, except by air, the impossibility of developing strategical pressure cancelled out this advantage. For similar reasons our administrative direction was nil, since, though our tactical resistance was strong, our strategical pressure was negligible.

### 5. Direction and the Human Element

I have now dealt in some detail with the organization of direction, and for a moment will turn from the mental and physical spheres to the moral. I do not intend to enter deeply into the moral side of direction, as I shall revert to this sphere of force when I examine the moral principles of war. It is of importance to remember, however, that moral force decides the degree of expression of physical force, and that, as the aim of direction is to expend force economically, the condition of moral force at the time of expenditure directly influences economy.

Many battles have been strategically and administratively

well founded, yet tactical results have been negligible, not because superiority of physical force was lacking, but because it lacked animation. Ultimately all depends upon what the man is willing to do, and the strength of a man's will depends very considerably on the absence of fear and fatigue. If administrative direction is wanting, discomfort results, and the will becomes personal in place of collective. When this happens a general's control weakens. If strategical direction is wanting, by degrees the men lose faith in their commander, and, through him, in their leaders; the stimulant of originality is wanting, no novelty of action magnifies his powers, the general appears to his men as one of themselves—a very ordinary person—and, as such, is subjected to criticism. If tactical direction is wanting, unnecessary losses result. It is not casualties in themselves which unnerve the men, for soldiers are not much stirred by the aspect of the dead, but what does unnerve them is *unremunerative* losses, lives foolishly thrown away; for then every time they see a dead man they say: "Another 'stiff 'un,' and what have we gained?" And this contemplation leads to another: "Perhaps we shall fill a similar billet to-morrow, and what for?" It is not death which demoralizes, but unnecessary death. The soldier will submit to any danger if led by a hero, but to few if led by a butcher. Once the soldier is only willing to fight because he fears military law more than he fears the enemy he ceases to be a reliable instrument.

### 6. The Principle of Determination

The plan is arrived at through an intellectual process of foreseeing, reasoning, and deciding, and before it can be transformed into the activity of war it must be given life. It is the general who verifies his plan by animating his instrument. This animation is governed by the principle of determination, and according to its application are the limits of the plan defined.

Throughout the history of war, courage, pluck, boldness, audacity, and determination have been terms employed to denote a quality which is of the utmost value both to the individual fighter, whether soldier or general, or to the army as a whole; yet historians have been content to accept it as a natural gift, and soldiers generally, especially in modern times, have followed suit. They have looked upon it as an element pure and simple, and have seldom attempted to analyse the influences of the conditions of war upon it, or to discover the nature of the relationships arising out of these influences.

Principles of war are not talismans, but abstract conceptions of general ideas. In themselves they possess no magical powers. It is useless to say: " I am determined to defeat the enemy "; for it is not the assertion which accomplishes defeat, but action. Direction is the resultant of three factors—concentration and distribution governed by economy of force; so is determination the resultant of three factors—originality—that is, action which will surprise and demoralize the enemy—and endurance governed by direction of force. What now are the moral conditions I am called upon to operate in? If I do this or that and the enemy does that or this, what moral conditions will arise, and how will these conditions influence my force, and through my force my plan, and through my plan my will? If my plan is destroyed, for the time being my will is paralysed, yet not one man may have been lost. The battle of Jena was not alone won on the heights of the Landgrafen Berg, but in the manœuvre which preceded their occupation; it was these manœuvres which demoralized Brunswick. It is in the conception quite as much as in the execution of an operation that success lies, and the link between conception and execution is animation—the moral tone of the instrument.

Like direction, determination is founded on knowledge, but, in particular, knowledge of a moral order. Through direction a general arranges his force, distributing and concentrating it, and forming it into an economic weapon; but through determination he controls its sentiments. If to direct the forces of war one man alone is required, so also is it solely within the province of one man to animate the instrument in its highest degree, and this one man is he who directs it.

I have already stated that committees and councils cannot govern armies, and though this fact is common knowledge, repeated in every text-book on war, in the last great war direction by committees and conferences was reduced to a fine art, an art in which the general in command became a constitutional monarch and the power which by right was his was relegated to his staff and delegated to his subordinates; it was a command by soviets.

The soldier, being utterly surprised by the magnitude of the war, and because the war was morally unlike anything he had expected, lost his mental equilibrium, and in herds of conferences sought to evade the responsibilities incumbent in determination by merging his powers of direction in the clatter of round-table talk. Had he realized what command meant, that command is a compound of autocracy and animation—that is, of deciding and of stimulating—he could not have acted as he did. It was because he was mentally fearful that he trusted in command by conference.

### 7. Jomini's Opinion on Councils of War

This lack in the individuality of command, which undoubtedly prolonged the war for months, is still a shibboleth in all armies. I make no apology, therefore, for the following long quotation from Jomini's *Art of War*; for whilst during peace-time soldiers are always talking about command, and the qualifications of the commander, the first thing they do when war is declared is to abrogate it. Jomini writes:

It has been thought, in succession, in almost all armies, that frequent councils of war, by aiding the commander with their advice, give more weight and effect to the direction of military operations. Doubtless if the commander were a Soubise, a Clermont, or a Mack, he might well find in a council of war opinions more valuable than his own; the majority of the opinions given might be preferable to his; but what success could be expected from operations conducted by others than those who have originated and arranged them? What must be the result of an operation which is but partially understood by the commander, since it is not his conception?

I have undergone a pitiable experience as prompter at headquarters, and no one has a better appreciation of the value of such services than myself, and it is particularly in a council of war that such a part is absurd. The greater the number and the higher the rank of the military officers who compose the council, the more difficult will it be to accomplish the triumph of truth and reason, however small be the amount of dissent.

What would have been the action of a council of war to which Napoleon proposed the movement of Arcola, the crossing of the Saint-Bernard, the manœuvre at Ulm, or that at Gera and Jena? The timid would have regarded them as rash, even to madness; others would have seen a thousand difficulties of execution, and all would have concurred in rejecting them; and if, on the contrary, they had been adopted, and had been executed by anyone but Napoleon, would they not certainly have proved failures?

In my opinion, councils of war are a deplorable resource, and can be useful only when concurring in opinion with the commander, in which case they may give him more confidence in his own judgment, and, in addition, assure him that his lieutenants, being of his opinion, will use every means to ensure the success of the movement. This is the only advantage of a council of war, which, moreover, should be simply consultative and have no further authority; but if, instead of this harmony, there should be difference of opinion, it can only produce unfortunate results.[1]

Before preparing his plan a general should tap all sources of information, including the local knowledge of his subordinates;

[1] *The Art of War*, p. 58.

then he prepares his plan, and finally issues it as an order, written or verbal. It is not for his subordinates to question it, but to carry it out. There should never be any great difficulty in this, if the intelligence services are efficient, but, as Clausewitz says: "The great difficulty is *to adhere steadfastly in execution to the principles which we have adopted*. . . . Therefore the free will, the mind of the general, finds itself impeded in its action at every instant, and it requires a peculiar strength of mind and understanding to overcome this resistance."[1] The will of the general, governed by his reason and imagination, is the directing and driving force of the plan. Smash this will, and the plan is smashed; weaken it, and the plan is weakened. The normal process of doing this is to attack the will of his subordinates, especially those in close contact with the troops; for these men do not see the state of the enemy, but their own state; "therefore the latter makes a much greater impression than the former, because in ordinary mortals *sensuous impressions* are more powerful than the *language of the understanding*."[2] Thus by disorganizing the combatants we demoralize their leaders, and by demoralizing the leaders we paralyse the will of their commander, which is the directing force of the battle, and which enables all parts of the instrument to co-operate.

### 8. The Animation of the Instrument

The conflict of reason and instinct is one of the outstanding problems of war. During peace-time all our efforts are directed to form an instrument which will react to its commander's will. In battle the organs of sensation are excited, and through the presence of danger fear is aroused, and the impulse resulting tends to a reaction from the will of the commander. He knows what is right, or at least is acting on an idea, and unless the will of his men responds to this idea their actions will be out of harmony with it. Thus the conflict is one between the self-assertion of the general and the self-preservation of his men, and unless, as I have shown in chapter vii., fear is balanced by *moral*, determination or the "encouraged will" cannot assert its power.

It is necessary for a moment to examine what is generally called freedom of will. Scientifically, will is only free when our volition is in agreement with cosmic laws and circumstances. In Nature all things move in certain ways. Water runs downhill, and is compelled to do so by the force of gravitation; there is no freedom of will about water. Man can, however, imagine water running uphill, and, if he does not understand the nature of

[1] *On War*, vol. iii., p. 222. [2] *Ibid.*, vol. iii., p. 226.

aqueous movements, he may try to make it run uphill. His failure is, however, preordained, and eventually, through trial and error, he learns that water will only flow downhill. When reason replaces trial and error, he discovers the reason why water will not flow uphill, and he calls his discovery the law of gravitation; and thus it is that, through this law and other laws, he discovers that freedom of will varies directly with his knowledge of the forces which govern the universe. *Complete obedience to the laws which control these forces is freedom of will*, and the closer this state is approached the freer is our will; and, conversely, the more distant we are from it the less free. Consequently our knowledge is the measure of our freedom, and if the idea a general wishes his men to carry out is a right idea, then it follows that, the more intelligent his men are, the more likely are they to carry it out economically, since intelligent men normally assimilate knowledge rapidly. If, however, the idea in question is a stupid one, then their intelligence will revolt against carrying it out. Men are not usually fatalists, for fatalism is freedom of will independent of conditions; in place, to intelligent men freedom of will is dependent upon conditions, and if they see that conditions are such that the idea cannot be carried out, then they may refuse to carry it out unless they have been informed of the reason why this infringement of the law of economy of force is required, or unless they have such implicit trust in the wisdom of their general that they realize that he is faced by a choice of two evils, and that, through their self-sacrifice, greater economy will finally result than through their self-preservation. Here we are confronted by several factors which control determination, the most important of which are related to the general. His knowledge and his prestige for doing right must be unimpeachable, and reliance in him must be so complete that the will of his men is merged into his own.

Though the physical loss resulting from disaster or defeat is obvious to all, and though moral loss, in so far as the endurance of the men is lowered, is frequently, though by no means always, recognized, what is seldom realized is that the main loss is in the will-power of the commander over his men. To him as an individual defeat means loss of prestige, which cannot be made good by reinforcements, or by rest and training, but only by success in the field. For a general to depend on disaster to teach him to be cunning means that his men must meanwhile endure the moral strain of war; in place, one who gains success at the lowest cost not only relieves this strain, but tempers the endurance of his troops. Very rightly did Roger Ascham say: "It is a costly wisdom that is bought by experience"; and equally wise was Benjamin Franklin when he wrote: "Experience is a dear school, but fools will learn in no other."

Information enables a general to know what to do; animation enables his men to carry out his orders with enthusiasm. Jackson accentuates this again and again in his book. Thus he says: "The human character is the subject of the military officer's study; for it is upon man that his trials are made. He must, therefore, know, in the most precise manner, what man can do, and what he cannot do; he must also know the means by which his exertions are to be animated to the utmost extent of exertion. The general's duty is consequently an arduous duty; the capacity of learning it is the gift of Nature; the school is in the camp and the cottage rather than in the city and the palace; for a man cannot know things in their foundations till he sees them without disguise; as he cannot judge of the hardships of service till he has felt them in experience. He may then judge of them correctly, and apply his rules without chance of incurring error."[1]

Throughout the whole course of history fear and love have been employed to animate armies, but, as Jackson truly says: "Fear and love are coverings; behind them must lurk the spirit of genius which cannot be fathomed; for, whether a commander be kind or severe, he cannot be great and prominent in the eye of the army unless he be admired for something unknown. It is thus that troops can only be properly animated by the superior and impenetrable genius of a commander, whose character stands before the army as a mirror, fixing the regards while it is bright and impenetrable, losing its virtue when its surface is soiled or softened so as to receive an impression. That a commander be a mirror, capable of animating an army, he must be impenetrable; but he cannot be impenetrable without possessing original genius. An original genius does not know his own powers. It thus commands attention, and it gives a covering of protection, in reality or idea, which proves a security against the impressions of fear."[2]

I have already pointed out that if a man does not possess original genius we cannot endow him with this quality, but genius is, after all, only exalted and spontaneous conformity to the law of economy of force, and, consequently, with the nine principles which emanate from this law; consequently the more we train ourselves to apply these principles correctly the nearer shall we approach equality with genius. A genius is possessed of a sublime freedom of will. This, as Jackson says, is a natural gift and a mystery to normal men; yet normal man himself can at least approach genius, if he cultivate a scientific freedom of

[1] *A Systematic View*, etc., p. 220. [2] *Ibid.*, p. 229.

will through obedience to the law of economy of force. To obey this law he must understand it, and for his will, which is law to his men, to be obeyed, they must understand him. All this is included in the principle of determination of force.

### 9. The Delegation of Responsibility

I now come to another and very important question which influences the application of the principle of determination, namely the delegation of the will to act. The general directs his army through his plan, and he animates it through the prestige he has created, and according to this direction and this animation is the force of his men expended. In ancient times the contact between direction, animation, and expenditure was immediate. Thus, for example, in the case of Alexander, he directed by his will, animated by his personal example, and expended the force of his army by word of command, because in his day the instrument was compact, closely articulated, and comparatively small. In short, the conditions of time and space were such as to permit of the personal appliance of the principle of determination. Yet even in his day he was compelled to delegate the command of his left wing to Parmenio, his second in command, reserving that of the right wing to himself. In fact, though he commanded his whole army, he only led the more important part; nevertheless, his command was close and his leadership of the right wing was intimate. If he had attempted to command both wings his leadership would have failed, since, being unable to judge conditions influencing the left wing at their true worth, he could not have determined their effects, and, consequently, could not have economically applied the principle of distribution.

This intimate control of the expenditure of force lasted until quite recent times; even as late as the battle of Waterloo we find Napoleon intimately commanding one side and Wellington the other. These generals have ceased to be leaders, but they are still in every sense commanders; in spite of the fact that Napoleon's commandership lacks the snap of youth, he is no longer what he was at Arcola and Jena.

To-day command has not only become divorced from leadership, but has become separated from the Napoleonic conception of commandership, which is that the general-in-chief commands his army in the same way as a craftsman commands his tools. He says: " In military operations I consult no one but myself."[1] Why? Because he himself only knew *exactly* what he wanted. And again: " In war, the first principle of the general-in-chief

[1] *Correspondance*, i., No. 339.

is to hide what he is doing, to see if he has the means to overcome all obstacles, and to do everything in his power to overcome them when he has made up his mind."[1]

To-day, if we are to accept the Great War of 1914–18 as our criterion, this conception of command has been replaced by one of delegation. This change-over first became generally apparent in 1866, and still more so in 1870. In the Russo-Japanese War of 1904–5 we find this same system in full play—a system which may be called the Prussian System, and which definitely introduces the modern epoch of war, though in truth it is not modern but very ancient, since Xerxes and Darius used it over two thousand years ago. It is not an evolution in the art of war, but a retrogression, placing, as it does, the determination of events in the structural order of battle rather than in the control of the instrument by one will. A plan was made and forces were deployed accordingly, command was delegated to the leaders of fractions, and, once the machine was set in motion, control over its direction became inanimate, for the machine moved forward compelled by brute strength and not guided by intelligence.

This system of command, based on the theory of brute force, led to the theory of superiority of numbers. An army a million strong would, like an avalanche, crush out of existence an army of but half its size. Initial direction was all-important; changes in this direction were anathema, since force of numbers would flatten out all obstacles, hence reserves were unimportant, for the cutting edge alone mattered, and if this edge consisted of six or seven men per yard of the enemy's entire frontier it could live on its own fat until the enemy was driven over the opposite frontier. This blind and monstrous theory of war reached its apex in 1914, and it failed ignominiously.

It is not here that I intend to examine its failure, but rather its results. In 1914 all nations saw it fail, but they could not see that one of the principal causes of its failure was the abrogation of the will of the general-in-chief as the determining factor in war. Right through the amorphous strugglings, surgings, flow and ebb of this blindest and most brutal of all wars—not brutal because of losses, but because of the lack of directing genius—we see no single general-in-chief fighting his own battle. In place, each formulates a plan and then delegates *his* responsibilities to *others*. The general-in-chief assumes the position of a chief of the staff, and his subordinates become commanders, and each battle is fought by a congeries of soviets—committees of generals who frequently rejoice over each other's defeats as full-heartedly as over their own successes. This system proved itself

[1] *Correspondance*, xx., No. 16372.

absurdly uneconomical, yet it is the system still accepted to-day!

In my opinion—and I have no two thoughts on this subject—a general-in-chief should *always* fight the main or decisive action himself, and should only delegate the direction of subordinate actions to subordinate commanders. Those under him have no corns that may not be trodden on, for the general-in-chief, in order to command, must be an absolute autocrat. How much of his plan he imparts to his subordinates depends on their personalities, on their will and their courage, and on how far their moral endurance permits of his ideas moulding their will. In them fear, or the absence of fear, is his guide. Sometimes a general-in-chief must keep his plan so secret that even during its execution he alone knows its full scope. For instance, if he plans a decisive attack which for success depends on a holding attack, and if this holding attack depends for success on the enemy considering it the decisive attack, he may be compelled, by the personality of the general to whom he has delegated the command of the holding attack, to withhold from him the true nature of the operation. If he tells him to attack as if it were a decisive attack, and then says: "Of course, it is not a decisive attack," his subordinate may lack the determination to attack full-heartedly.

Thus we see that delegation of command is not so simple a problem as it appeared in the Great War, for it is a problem of psychology and not of arithmetic. No general-in-chief purposely wants to keep his subordinates in the dark, but circumstances sometimes compel him to do so. To treat all men as equal is to reduce human nature to a mechanical principle; a general-in-chief is not a Communist, save perhaps in bellicosity. No two men are alike; what, then, are their differences? For on these conditions depends how we determine the delegation of our responsibilities.

Even when command is delegated, direction over its determination can frequently be maintained. Thus, to revert to the above example of the holding attack, the general-in-chief does not really want it to be driven to a conclusion, for all he wants it to do is to bite and hold on to the enemy, to fix him in a position in which he can be annihilated by another force. Yet he orders his subordinate to attack in full. How can he control this attack? By allotting to it a force which cannot do more than hold. He, in fact, determines the endurance, in this case physical, of the attack by a just distribution of force, and yet he may not tell his subordinates that he is doing this.

In the last great war delegation of responsibility was stimulated by the promotion of mediocrity to command. The higher the command, normally, the less efficient became the general. Men rose in rank according to the date of their birth or of their commissions, and seldom because they possessed ability. Senility sat heavy on all armies, since it is the exception and not the rule that old men prove the best commanders; and history proves this again and again. Napoleon said: "It is at night-time that a general-in-chief should work; if he tires himself uselessly during the day, his fatigue will overcome him in the evening. At Vittoria we were beaten because Joseph slept too much. If I had slept on the night of the battle of Eckmühl I should never have carried out that superb manœuvre, the finest I ever accomplished. I multiplied myself by my activity . . . a general-in-chief should not sleep."[1] In place, what do we see? Elderly men sleeping soundly, unruffled even by hopes of success or dreams of failure, for they have delegated all responsibility to others, save that of the heavy guns and the rearmost transport lines. They can determine nothing, so they slumber; and how can one blame them? Such was command during 1914–18, a command which would have made Darius blush.

## 10. The Meaning of Initiative

Having examined the problems of the animation of the instrument and the delegation of responsibility, I come to another problem of equal importance, namely, the problem of initiative, for action depends largely on this quality—the will to act.

Throughout a great battle, a campaign, and a war, the principle of direction is maintained by correct concentration and distribution, and merges into the principle of determination when moral endurance is proof to withstand surprise, and, be it remembered, nearly every change in the conditions of war results in an unexpected situation, or one which demands an alteration in action, and, consequently, in the determination of will. On the part of the general-in-chief this alteration may prove extremely difficult, unless he has foreseen its likelihood and has distributed and concentrated his troops accordingly, for his main source of initiative lies in his reserves. With his subordinate commanders and with the leaders of the men the problem is more difficult, for, though they should also maintain reserves in order to meet unexpected situations, they do not possess the same freedom over distribution as the general-in-chief.

If subordinate commanders have definitely been delegated

[1] *Sainte-Hélène, Journal inédit*, Général Gourgaud, ii., p. 159.

control of certain operations, then they should be allowed full freedom of action within the terms of references of the plan. In such cases the interference of the general-in-chief is illegitimate, since delegation carries with it responsibility, and responsibility can only economically be centred in the will of one man. Without this centralization of will true initiative becomes impossible.

In the case of leaders—that is, of officers serving under a commander—their initiative depends on their ability to determine the true values of changes in conditions with reference to the endurance of their men. How far has each change reduced or increased this endurance, how far has it effected surprise and consequent demoralization—actual or potential—and how far has it stimulated the fighting spirit of the troops? It is the balance between the principles of surprise and endurance which results in determination, and it is the principle of determination which sets a limit to movement.

Once the leader has thought out these changes his action cannot solely be determined by what is of immediate benefit to his troops, but it must be referred back to the original plan and directed accordingly.

If, in the opinion of the leader, the plan has, through change in conditions, become inoperative, then he ceases to be a leader, and becomes, for the time being, an independent commander, and he must act as if he were a general-in-chief. That is to say, he must replace the inoperative plan by an operative one—that is, one which will permit of the economical expenditure of force. To carry on a plan which manifestly has failed is the act of a fool, whether he be the general-in-chief or a private soldier. Once again we come back to our starting-point, namely, intelligence.

### II. Singleness of Purpose

Singleness of purpose and simplicity of organization are powerful means of enabling determination to express itself. The old Roman saying that a nation should not wage two wars simultaneously is a wise one, and neither should a general. In the Great War of 1914–18, amongst ourselves, we see the commander-in-chief in France not so much commanding the British armies as waging war with the Government at home. His back is to the enemy, and he faces those whom politeness demands should be called his friends. In chapter v. I suggested a means of overcoming this difficulty, namely the appointment of a generalissimo who possesses singleness of purpose towards fixing the military object of the war. Policy must be clean cut, for on its stability depends the solidarity of the forces with which it is proposed to gain the military object, the gaining of which psychologically depends on the endurance of the " will to win."

This will should be centred in the mind of the general-in-chief, whose plan of action expresses the military method of enforcing the national policy. This plan must also be clean cut; that is to say, it must be so simple that it contains no undetermined or undeterminable complexities.

As the stability of this plan will depend on the stability of the policy, the commander-in-chief must not only be acquainted with the nature of this policy, but with any changes rendered necessary in it due to fluctuations in national and international conditions. Inversely, any important changes in plan will entail modifications in policy, consequently we find that both the plan and the policy are correlatives, since there exists the closest relationship between them, their respective values being determined by each other's stability.

As every policy must be plastic enough to admit of fluctuations in national conditions, such as commerce, industry, social solidarity, and neutral and hostile influences, so must every plan be plastic enough to take the impressions of war; that is, a plan must be so thought out that it is possible to change its shape without cracking its substance.

This plasticity is determined, psychologically, by the degrees of mentality possessed by the two opposing forces. There is the determination between the two commanders-in-chief and between them and their men, and ultimately between the two forces of men themselves. The " will to win " is, therefore, first a duel between two brains, each controlling a weapon called an army; and, secondly, a struggle between two armies, each equipped with various means of waging war. If all the various weapons, each influencing in its own degree the mentality of the wielder and that of his opponent, can be reduced in numbers, the principle of determination becomes more simple of application. If, again, similarity of protection is possible, it becomes simpler still. And if, finally, similarity of movement be added, physically the simplest form of army is evolved.

If the will and *moral* of each individual can be brought to a high but equal level, and his fear to a low and equal level, the commander-in-chief will possess known quantities out of which to construct his plan. We find, therefore, that, in its broadest sense, the principle of determination aims at obtaining a rational simplification of the means, so that the will of both the chief and his men may be directed towards the objective, and concentrated on it.

### 12. The Principle of Mobility

Mobility is the third controlling principle of war, a principle which endows all military operations with activity, whether offensive, protective, or logistical, and it finds its expression through the element of movement which draws its power from physical energy. Mobility is, therefore, the principle which governs the expenditure of force, and, as I stated in the last chapter, if it were possible to move correctly, then this principle would coincide with the law of economy of force.

In chapter viii. I examined movement in its forms of the approach and attack, or, in other words, protective and offensive movements. Though the former constitutes the base of strategy and the latter of tactics, there is no definite dividing-line between these two. Strategy cannot be divorced from tactics, for, in the battle itself, strategical movements are continued in the form of the approach. To state that strategy comprises all movements before battle and tactics all movements during battle is to suppose that a division between these two essentials can be established by the firing of a shot. Further, it is apt to suggest that the principle which governs strategical movement is not the same as the one which governs tactical movement; consequently that in place of one principle there are two. The difference is not to be sought in the principle of mobility, but in the conditions in which it is applied. These conditions, if rightly read, dictate which elements of war should become the predominant partner, and, according as one element becomes paramount, so does mobility change its form. Thus, if conditions enable movement to take place without the use of weapons, the form which mobility takes is strategical, whether during, or before, or after battle. Or, again, if weapons have to be used to facilitate movement, then the form is tactical. I mention this here because the dependence of the principles of war on the elements of war as influenced by the conditions of war, which either resist or facilitate movement, must never be overlooked.

### 13. The Dependence of Mobility on the Conditions of War

Having provisionally decided upon our objective, and having distributed our forces protectively and offensively, the next question to decide is how to move them, and it is here that a close study of the physical conditions of war come to our assistance. Of these, ground and communications are of the highest importance, and, though this is obvious, it is frequently overlooked.

Throughout history, rivers have constituted the main lines of communication, and even to-day it is along the river valleys that the greater number of roads and railways wind their way. All these communications lead to and from towns, which become centres of communications, and, consequently, positions of strategical importance. Along rivers, where the soil is usually alluvial, cultivation is profitable, and during war-time cultivated areas constitute an administrative assistant and a tactical resistant, that is to say, they assist the supplying of armies but impede their movement on and off the battlefield.

We see, therefore, that communications which follow the river-lines influence in varying degree the strategical, tactical, and administrative movements of armies. Thus certain roads and railways have to be followed, consequently approaches cannot be kept secret; and as these roads and railways often run along low ground commanded by high, and through towns which can be converted into field fortresses, and through cultivated country which provides these strong points with all types of obstacles to their approach—hedges, ploughed fields, plantations, crops, ditches, wired fences, and isolated houses, etc.—the defender has much to support him in holding them, and the attacker much to overcome in advancing through them. All these conditions must be carefully weighed before the principle of mobility can be applied.

Great wars, normally, take place in well-watered areas, for these, being generally the centres of civilization, not only offer economic objectives, but give rise to economic and political disputes. On the other hand, small wars generally take place in badly watered districts—mountainous and desert country where natural obstacles abound. In these areas wars are waged more against these obstacles than against the enemy himself, and, communications being scanty and difficult to protect, supply usually takes precedence over tactics.

In the past, in both types of war, communications, their defence and attack, have constituted the woof and warp of military operations. In mechanical warfare our present theory of communications will have to be modified. In great wars—that is, wars in which battles are fought on flat and undulating ground—the width of roads will be widened indefinitely until they cover vast areas, and possibly entire theatres of operations. In desert warfare the same will occur; but in mountain warfare, though precipitous valleys will restrict lateral movement, the roads and tracks following them will be rendered far less vulnerable to flank attack by the use of armoured mechanical supply columns. I note this here for the possibilities of mechanical warfare must to-day be considered when we study the conditions of an area of operations with reference to the principle of mobility.

### 14. The Dependence of Mobility on the Principles of War

I have already stressed the point that it is not possible correctly to apply any one of the principles of war without reference to the remainder. In the present case this becomes readily apparent. If the objective selected cannot be approached, the principle of direction is violated, because the principle of mobility cannot be applied. If communications lead to an impossible offensive area, then, if we follow them, we shall violate the principle of mobility through rendering ourselves powerless to apply the offensive and incidentally violate the principle of concentration. And, if they lead through areas which cannot be protected by the means at our disposal, then again shall we violate the principle of mobility by being unable to apply that of security, and without security our distribution has proved itself faulty. We see, therefore, that the line of least resistance is not necessarily the easiest line to advance by, but, in place, the line which will enable protected offensive action to succeed.

To apply the principle of mobility we must have a definite object as the directing idea of movement. The danger in changing an object mainly lies in the changes of movements which result, and especially of administrative movements. To take a very simple case: a battalion is drawn up in line on its parade ground, with its transport in rear of it. It is facing east, when an order is given for it to face west. As regards the men, all that is necessary is to say, "About turn," but the transport has to move to its new position either by going round the battalion or through it. When armies are concerned, such an operation is normally impossible, and even lesser degrees of change of direction generally lead to friction, and consequent loss of energy.

To maintain the principle of mobility, not only must the objective be fixed, but the base of operations must be secured as well as the lines of communication running forward from this base. A change of base is even more dangerous than a change of objective. It is for this reason that attacks on communications, rather than against the armies themselves, form the most important operations in war. An attack against the base of an army frequently forces a commander to change his object. It has therefore a dual influence; it not only forces an enemy to change his intention, but to fight for the maintenance of his communications in place of attempting to destroy his adversary. A good example of such an operation is the opening campaign of 1914. The French base of operations was Paris, and the French object was an offensive in Lorraine. The movement of the German right wing through Belgium against the French communications caused General Joffre to abandon his plan in order to secure his base; it also forced Sir John French to change his base from Havre to St. Nazaire.

The principle of mobility, we see, is immediately dependent on the principles of security and of offensive action. As these two principles are maintained, so does mobility flourish, and as they are violated, so does it wither away. In their turn, security and offensive action are determined by the state of moral endurance, or of demoralization, existing in the troops themselves, which is dependent on the correctness of distribution and concentration as expressed in the direction of the operation. If direction is, or rather could be, perfect, then the law of economy of force has been obeyed. Obedience to this law does not in itself guarantee victory, but what it does guarantee is the most profitable expenditure of force in the circumstances which surround it.

## 15. The Movement of Ideas

The expenditure of physical force through movement is, as I have shown, dependent on the will to move, and its economical expenditure on the direction of this will. The first is generally recognized, but, though the second is recognized in so far that every sane man knows that the right way is better than the wrong way, amongst soldiers so little is known of the science of movement that the art of moving is considered the natural prerogative of each separate individual. Hence, when a new idea is put forward, in place of it being analysed and valued it normally is accepted or rejected, not on sufficient evidence, but on personal predilection. I intend, therefore, first of all to examine the movement of ideas, and, secondly, the existing organization of movement, for, in my opinion, the changes which to-day face all armies are mainly connected with movement, and, unless ideas are scientifically examined, organization will remain unchanged, or the changes introduced will be uneconomical.

Movement of ideas depends on liberty of thought, just as movement of things depends on liberty of action, and unless ideas—strategical, tactical, and administrative—are permitted to move, concentration of effort will not result, and in proportion as unity of action is lacking, so will the moral and physical strength of an army be squandered in detail until a time arrives in which the minimum result is obtained from the maximum effort.

The central idea of an army is known as its doctrine, which to be sound must be based on the principles of war, and which to be effective must be elastic enough to admit of mutation in accordance with change in circumstances. In its ultimate relationship to the human understanding this central idea or doctrine is nothing else than common sense—that is, action adapted to circumstances. In itself, the danger of a doctrine is that it is apt to ossify into a dogma, and to be seized upon by mental emasculates who lack virility of judgment, and who are only too grateful to rest assured that their actions, however inept, find justification in a book, which, if they think at all, is, in their opinion, written in order to exonerate them from doing so. In the past many armies have been destroyed by internal discord, and some have been destroyed by the weapons of their antagonists, but the majority have perished through adhering to dogmas springing from their past successes—that is, self-destruction or suicide through inertia of mind.

Mental lassitude, or the abiding by the letter in place of the spirit of the law, which so frequently passes for military ability, is the dry rot, not only of armies, but of kingdoms, republics, and empires.

Though an army should operate according to the idea which, through methodical training, has become part of its nature, the brain of a commander must in no way be hampered by preconceived or fixed opinions; for, whilst it is right that the soldier should have absolute confidence in himself and his comrades, and through this confidence should consider himself invincible, it is never right that the commander should consider himself undefeatable. Contempt for an enemy, however badly led, has frequently led to disaster. It is, therefore, the first duty of a commander to maintain his doctrine in solution, so that it may easily take the mould of whatever circumstances it may have to be cast in.

We here obtain a dual conception of doctrine. In the first case, doctrine must be looked upon as a fixed method of procedure, so that, when an order is issued, all may understand it, and unity of action may result. In the second case, doctrine must be looked upon as power to formulate a correct judgment of circumstances and to devise a course of procedure which will fit conditions. If this be a correct definition, then it stands to reason that, if the will of the commander is to control the actions of his army, the doctrine of an army must be such as will permit of *any* rational idea moving it without friction. The question now arises: How can we train our men to follow a method which will in no way hamper the liberty of thought of their commander? The answer is: By basing the art of war on the science of war. If this be done, then the commander who thinks scientifically will find at his disposal an instrument on which, metaphorically, he can at will play any tune. This means that, until a science of war has been formulated, it is not possible to establish a doctrine which can be other than transient. In the past, practically every doctrine established during peace-time has proved itself to be obsolescent immediately it is put to the test of war; the reason being that these doctrines have been built on rules of strategical and tactical procedure dependent on the success or failure of fixed organizations, such as a battalion of infantry, a regiment of cavalry, etc., in varying circumstances, in place of on the elements of war. I will now attempt to explain this more fully by examining the organization of military movement.

### 16. The Organization of Movement

In chapter v. I examined at some length the structure of an army, and, in brief, I stated that formerly, and even to-day, tactical organization was based on the following idea: whilst the guns protect the infantry, the infantry attack the enemy's infantry, and when the enemy is demoralized, the cavalry charge home and annihilate him. If we examine this idea we shall see that:

(i.) Infantry are related to offensive power, and that the more this power is protected the stronger it will be.

(ii.) Artillery are related to protection, and the more it can protect the infantry the more will their power be economized.

(iii.) The cavalry are related to movement, and the more thoroughly the infantry carry out their work the sooner will the cavalry be able to operate.

Briefly, the gun protects rifle-power in order that mobility may be attained by the cavalry. Formerly cavalry was the decisive arm, but to-day it is no longer so, and as infantry, when pursuing, cannot move faster than the retreating enemy, the result is that pursuits have become less and less frequent. Throughout the war, on the Western Front, there were many retirements and advances, but not a single sustained pursuit on an important scale. In Palestine a magnificent pursuit was carried out in the autumn of 1918, because conditions favoured cavalry movement. The crucial question in the modern attack is: How to re-establish mobility by ability to pursue—that is, how to annihilate the resistance of the enemy?

An answer to this question I feel can be found in the tank, which, being able to move faster than infantry, *can* pursue, and not only pursue, but also attack through virtue of its armour. As the tank can use its weapons and carry its own protection when in movement, it will enable the present static fighting to be replaced by dynamic fighting; that is to say, the soldier, whether infantryman or gunner, will not have to halt in order to deliver blows, but will do so whilst in movement. This possibility must sooner or later lead to a radical recasting of tactical organization, as radical as that which followed the introduction of gunpowder. Yet the anatomy of whatever organization replaces the existing one will be in nature the same, for it must be based on the elements of war. Thus, if we examine history we shall always find that when tactics flourished there were three classes of fighters, namely offensive or close-combat troops, protective or distant fighting troops, and mobile or pursuit troops. Whenever one of these classes disappeared, such as I have noted was the case during the Middle Ages, tactics declined, and the art of war grew primitive in nature. To-day we are entering a new epoch of war, and if our tactics are to be maintained at a high level we shall have to reorganize our forces according to the changed values of the three physical elements of war so that the mind of the commander may control the battle; for unless he can control it he cannot apply the principle of mobility. In other words, he must so organize his forces that this principle can be applied in its fullest extent—that is, with the least possible loss of energy through friction or delay.

### 17. The Endurance of Mobility in War

Once we have created an organization which will enable movement to find full expression, the next problem to solve is the maintenance of movement during active operations. If an army be compared to a machine which draws its power from a series of accumulators, then, if its commander wishes to maintain movement, he can only do so by refilling one set of accumulators while the other set is in the process of being exhausted.

In war the power to move must first be considered as the general will to move. In battle the forward impulse comes from the leaders and the troops themselves; they are, in fact, self-propelling projectiles, and are not impelled forward by the explosive energy of command. Such energy scarcely, if ever, exists, but what does exist is direction to its impulse and the reinforcing or recharging of this impulse with more power by means of reserves. Reserves not only endow the combatants with physical energy, but with moral power and security which impel them forward.

In the initial phases of a war it may be laid down as a general maxim that reserves cannot be too strong, and in these phases, when conditions and intentions are still uncertain, the principle of mobility is normally maintained rather by possessing power to move in its potential form—that is, locked up in a large reserve —than in its active form of an extensive or intensive offensive.

In war, reserves form the capital of the commander, and, if he opens the game with a maximum stake, it may not be long before he finds himself bankrupt. A good player knows the value of a cautious game until he can judge the value of his opponent's skill, and then the value of an audacious use of his capital. In war it is the same. Maintaining the initiative does not necessarily mean attacking and advancing. If the reserves be strong, it may frequently mean defending and retiring in order to create a situation in which their use may lead to decisive victory. In prolonged actions, as the original reserves are used up so must fresh ones be created in order to maintain power to move, and through movement influence the battle.

## 18. The Influence of Movement on Doctrine

Earlier in this chapter, by examining the conditions of communications, I pointed out the influence of the conditions of war on the application of the principle of mobility. Sometimes conditions are so adverse that it is most difficult to apply a principle, and if any one principle cannot be applied, then all the remaining principles must suffer.

Before the outbreak of the Great War all civilized armies were imbued with the spirit of the offensive, and simultaneously they were equipped with weapons of great power, such as the magazine rifle, the machine-gun, and the quick-firing field-gun. War was not thought of in terms of security; in fact, the application of the principle of security to the changes introduced by these weapons was grossly neglected. The result was that within a few weeks of the declaration of hostilities movement ceased, because conditions were such that the principle of mobility could not be applied with the existing instrument.

Now, it is beyond question an axiom that nothing can be

accomplished without movement, and it is a self-evident fact that a principle is not a means of war, but an abstract idea which, when translated into action, directs the use of the means employed. I have already stressed the fact that these means are changing. Before the war the main changes were towards an increase in weapon power; to-day they veer towards an increase of movement. We must understand this change, for, if we do not, we shall never learn how to apply the principle of mobility when the next war is declared.

In 1913 we did not realize the protective power of weapons, and the result was static warfare. In the next war, if we do not realize the influence of new forms of movement on weapons and protection, the war, in place of being in nature static, will be dynamic in the extreme; we shall be swept into the sea or into some neutral country.

To-day our conception of strategical, logistical, and administrative movements is what I have called one-dimensional. In a few years' time, when armies largely consist of tracked vehicles and aircraft, to this one-dimensional movement will be added movement in two or three dimensions. I do not here intend to speculate as to the nature of these changes; visibly they will be immense. In place, I wish to emphasize this fact that, unless we are willing to scrap our old conceptions of war and replace them by new ones, when war comes we shall without doubt attempt to apply the principle of mobility to new conditions as if they were old conditions, and without doubt we shall be surprised by our ignorance. Unless conditions are understood it is not possible, save by chance, to apply a principle correctly. We do not know these conditions; nevertheless, by making use of our intelligence we can discover their tendencies. We can test out ideas concerning them, and so gain experiences through a process of trial and error. It is for this reason that in place of considering mobility from the normal and stereotyped point of view—of interior and exterior lines of movement and manœuvre, and of parallel, oblique, eccentric, and concentric marches—I have not only dealt with the relationship of the principle of mobility to the elements, conditions, and remaining principles of war, but have also discussed the movement of ideas.

Armies are conservative organizations; they adapt themselves slowly to new environments, and especially to new mental surroundings. To-day a new epoch of war is dawning, and we are surrounded by a veritable fog of new ideas. We must neither accept them as they stand nor pass them by, but we *must* examine them and *test* out their values. What are they, and what changes do they foretell? If armies are to be endowed with a new means of movement, then most of the existing offensive and protective means of waging war will be changed. As the three physical elements of war change their present values, so must our present conception of war—the expression and value of the mental elements—change with them and not only with them, but we must foresee these changes. If mentally we cannot keep pace with the changes in the physical elements of war—the changes in weapons, movement, and protection—then our strategy and tactics will remain obsolete; that is to say, they will not enable us to express the principles of war when once again we are called upon to apply them. We shall go to war as we did in 1914—under a misconception. If fortune favours us on the battlefields, we shall learn from the changed nature of these elements most costly lessons. If our luck be out, or if our adversary be mentally superior to ourselves, we shall be annihilated, because whilst in 1914 we misjudged weapons—weapons which could be countered by the use of trenches—in the next war we shall have misjudged movement, which has rightly been called "the soul of war."

# CHAPTER XIII

## THE PRINCIPLES OF PRESSURE

The first rule of practice is to do all things at the right time and in their proper place; to proportion the means to the ends and the ends to the means; above all, to know what is possible, and to confine one's endeavours within the limits of the feasible.—J. J. SYLVESTER.

### 1. THE PRINCIPLE OF CONCENTRATION

THE relationship of the three controlling principles of war to the remaining six may be expressed as follows:

(i.) The principle of direction works through that of distribution by means of concentration.
(ii.) The principle of determination works through that of endurance by means of surprise.
(iii.) And the principle of mobility works through that of security by means of offensive action.

Though the three controlling principles are the resultant of the three principles of pressure and the three of resistance, they nevertheless direct, determine, and move the elements from which these principles originate; three being active and three being stable, the former being based on the latter. As controlled activity is our aim, for nothing can be economically attained in war by pure resistance, I will examine the principles of pressure first; these are the principles of concentration, surprise, and offensive action.

Clausewitz, when considering the "Plan of War when the Destruction of the Enemy is the Object," declares that "two fundamental principles reign throughout the whole plan of the war, and serve as a guide for everything else.

"The first is: to reduce the weight of the enemy's power into as few centres of gravity as possible, into one if it can be done; again, to confine the attack against these centres of force to as few principal undertakings as possible, and one if possible; lastly, to keep all secondary undertakings as subordinate as possible. In a word, the first principle is *to concentrate as much as possible.*

"The second principle runs thus—*to act as swiftly as possible*; therefore to allow of no delay or detour without sufficient reason." [1]

Clausewitz drew most of his ideas from a close study of the Napoleonic wars, in which, again and again, he saw the Emperor applying the principle of concentration. In the *Correspondance de Napoléon*, again and again, may be read sentences such as the following:

Your army is too dispersed; it should always march in such a manner as to be able in a single day to unite on any one battlefield. With 15,000 men I could beat your 36,000.[2]

My intention is to concentrate all my forces on my extreme right . . .. in such a way as to have nearly 200,000 men concentrated on the same battlefield.[3]

There are systems of waging war just as there are of carrying out sieges. Concentrate fire against a single point, and once the breach is made equilibrium is broken, all action becomes useless, and the place is taken.[4]

For concentration of force to be effected with rapidity, the framework of every plan must be extremely elastic, since conditions are always changing, and our knowledge of them is generally so limited that a large margin must be left over for the unexpected; consequently concentration of force is closely related, not only to distribution and direction of force, but to endurance and surprise.

Once our object has been decided on and the direction towards our objective fixed, the next question is to concentrate force against this objective—that is, to seek a decision.

If we decide that we can securely concentrate superiority of force against the decisive point, then our concentration will normally follow the line of greatest traction, as the initiative is ours; but, if security is doubtful, then we must decide between this line and the line of least resistance—that is, the line along which opposition will be weakest; if, however, superiority be deficient, we must create a line of greatest traction, or of least resistance, by manœuvre or surprise.

In the first case, the condition which governs the line of greatest traction is our own distribution; in the second case, the condition which governs the line of least resistance is the enemy's distribution, and in the third, the main condition is the relationship between our own and the enemy's distribution. Finally, the line we should choose is the one which will enable us to attain our object with the highest economy of force.

[1] *On War*, vol. iii., pp. 140, 141. [2] *Correspondance*, xii., No. 9808.
[3] *Ibid.*, xiii., No. 10920. [4] *Correspondance, inédite*, 13th July, 1794.

Once we have decided where to effect our concentration and how to secure its movement by correct distribution of force, our next problem is the organization of our force for offensive action of a decisive character. We must give it structure, and see that this structure can be maintained and controlled. Here the solution centres very largely round the strength of our reserves.

Having organized our hammer-head, we must next see that the moral and physical forces which wield the weapon are so expended that endurance is maintained ; this demands a detailed examination of the conditions in which expenditure of force will take place.

I will now examine a few, and only a few, of the many aspects of this principle.

## 2. The Direction of Concentration

Of all the principles of war the best known is probably that of concentration of force, and yet it is one which is constantly being neglected or misapplied. One of the reasons for this is that, though during peace-time military conditions are studied, the reality of war is forgotten, and directly war is declared this reality manifests as a fog which obscures or distorts actualities to such an extent that mental balance is lost, and without this balance concentration is most difficult to establish.

If we intend to concentrate a force of men, we must first know where the men are, and, secondly, the place at which we intend to mass them. As we generally know where they are, the second question is the only one which need be considered. Where, then, should we mass them ?

This question cannot be answered off-hand. We cannot always, like von Moltke, say : " Direction, Paris ; objective, the enemy wherever met," unless we know that by advancing on Paris the enemy will place himself between Paris and ourselves. We cannot know this for certain, but there are many conditions which we can know—such as the nature of the theatre of war ; the system of communications traversing it ; the fortresses securing these communications ; the commercial and industrial centres ; and a host of other factors. From these we can plan out a strategical and tactical map on the lines of a geological chart, and from this map we can learn the possible and then the probable movements of the enemy.

As we seldom can take it for granted that the enemy will adopt one definite course of action, we cannot concentrate our forces against one sector of a given front, therefore at the commencement of a campaign, however offensive may be our intentions, without losing freedom of movement, we should hold as large a reserve as possible in hand. This reserve we should cover and secure by a screen of strategic forces or advanced guards, the duty of which is to discover the enemy. Once contact is gained, then can our strategical plan be developed into a tactical one.

Directly the battle area has been selected, concentration begins with the application of the principle of distribution of force. The area is divided into defensive and offensive zones. In the former the idea is to resist attack, and in the latter to deliver it. By applying distribution of force we settle the question of bulk numbers, and, once the bulk we have allotted for offensive action has been decided on, the next step is to distribute it in such a manner that concentration of force is attained or attainable at the decisive point.

Leaving subsidiary operations out of the question, we first of all select our decisive point of attack, and then plan our main attack with a view to assist us in gaining this point. The object of the main attack is not to seize this point, but to prepare the way for a fresh body of troops to do so. The main attack must, therefore, through offensive action, force the enemy to draw on his reserves, so that freedom of action may be gained for our own reserves. Frequently it happens that we are unable to select a decisive point before engaging the enemy—in this case, power to apply the principle of concentration must be drawn from the same source, namely the reserves. The more it becomes necessary to fight for information, paradoxical as it may seem, the stronger must our reserves be. Consequently, if seeking information, through offensive action, demands so great an expenditure of force as to lead to a depletion of the reserves, more often than not, the wiser course is to assume a defensive attitude and let the enemy attack. Though this may mean that the enemy will push our defensive forces back, it does not necessarily mean that by so doing he has gained the initiative, for the initiative lies in the potential strengths of the reserves, and he who possesses the strongest reserves, as long as they are well placed, is master of this deciding force.

### 3. The Relationship of Concentration to Reserve Force

In the application of the principle of concentration a frequent mistake is to mass offensive forces against a selected point when it is impossible to surprise this point. This mistake originates in failure to appreciate that concentration, in nine cases out of ten, means *keeping* troops *out of battle,* and not thrusting them in. Men are not machines, and even machines require periods of rest and overhauling. Men have a limited physical endurance, and it is this endurance which must be economized. If 10,000 men attack a position simultaneously, the majority of these men will be exhausted simultaneously. If 6,000 men attack, and 4,000 are held in reserve, even if the enemy numbers 10,000, by the time the energy of the 6,000 is exhausted, that of the 4,000 in reserve will, in all probability, be greater than the residual energy of the enemy—that is to say, if he has employed the whole of his forces in the attack. In practice, as well as in theory, reserves can seldom be too strong. Again, the supply of reserves must be continuous, by which I mean that at no time during a battle or campaign should a reserve force be entirely used up. This means that directly a commander is compelled to draw on his reserves he should simultaneously withdraw exhausted troops to take their place. As the recuperation of these troops will depend on the residual energy possessed by them at the time of withdrawal, unless the original reserves are exceedingly strong, these troops should be withdrawn before their endurance, especially moral endurance, is exhausted. It follows, therefore, that the true psychological moment to withdraw troops into reserve is immediately after they have gained a success, and not when they are so used up that failure stares them in the face.

A general should always remember that a shattered front may demoralize an intact rear. Conversely, a victorious front, if it be withdrawn into reserve, will act as a moralizing tonic to every man behind it. If men are withdrawn into reserve with their tails well over their backs, all drooping tails in rear will assume a like attitude. To squeeze men like lemons, and then place them in reserve, is the act of a criminal lunatic.

From this brief survey of the value of reserve force as the foundation of concentration, it will be seen that it is impossible economically to allot frontages of attack without reference to reserves. In building up an offensive plan, directly security has been established, the next question to settle is the strength of the reserves. If the area of the decisive attack can be settled beforehand, this problem is not a difficult one, but if it cannot be, then every man who can be held in reserve should be held in reserve, and, unless the enemy is decidedly inferior to oneself, or unless surprise can be effected, if the reserve forces fall much below half the total forces, a commander should consider twice whether he will attack or not.

### 4. Concentration and Strategical Distribution

To distribute troops strategically has nothing whatever to do with mobilizing them in certain areas, but in so placing them in the theatre of war that they simultaneously can maintain freedom of movement and compel the enemy to conform to the plan adopted.

The most certain way of influencing the enemy is to threaten his communications in such a manner that he is forced to fight on a front parallel in place of at right-angles to them. Sometimes (as was the case at the battle of Jena) it is possible so completely to out-manœuvre an enemy that he finds himself facing his base. In both cases the initial attack delivered is in nature a moral one. This is the true act of attrition which should precede the decisive attack. Its supreme value lies in the fact that the enemy is being demoralized by manœuvre in place of by attack, consequently the whole of the forces engaged may be held in reserve.

From this we see the extreme importance of the initial strategical distribution on concentration of force, and that the application of force does not necessarily mean physical force but moral force, and that the greater the moral pressure we use in war the less need be the physical force we concentrate for the decisive battle. Between these two forces there is a radical difference; for, whilst expenditure of physical force leads to a loss of endurance, the moral attack on an enemy, by forcing him to conform to our will, enhances in place of reduces the moral of our men. The moral attack has, therefore, this immense advantage, the more it succeeds the higher becomes our moral power. The maintenance of the initiative does not, therefore, lie so much in physically destroying the enemy as in reducing him to a moral wreck. The most potent form of concentration is, consequently, the strategical surprise.

### 5. The Nature of the Force Concentrated

Before concentration is arranged for it is as well to decide upon the nature of the force to be applied. Concentration, from the point of view of battle, has for centuries been based on the maxim of "superiority of *numbers* at the decisive point," because numbers were the co-efficient of weapons, each man normally being a one-weapon mounting. As a general rule, this maxim no longer holds good, and in its place must be substituted "superiority of weapons, means of protection, and movement." Men, in themselves, are an encumbrance on the battlefield, and the fewer we employ, without detracting from our weapon-power, the greater will be our concentration of strength. If the area in which a decision is to be sought is held by hostile infantry, to concentrate masses of infantry against them, when we can concentrate tanks, is to violate this principle. If this area is, however, totally unsuited to tank action, as was the case in Flanders in August 1917, a violation will equally occur it we employ them. If the enemy's communications run through a defile, and we can attack these communications by aircraft, it is useless battering ourselves to pieces against the enemy's front. From these examples the point I wish to accentuate is that as conditions vary, so does the application of the principle of concentration differ. It demands selection of force as well as mass, and suitability of force as well as numbers. Like every other principle, it must be applied according to conditions; it cannot be applied by rule, and it cannot safely be applied unless the remaining principles assist in its application.

### 6. The Position of the Force Concentrated

The first step in attaining tactical concentration is to deny freedom of movement to the enemy. This can be accomplished either by manœuvre, or by definitely halting him, or by forcing him to deploy, and, whilst he is deploying, to attack him. To effect a tactical concentration it is first necessary to hold and then to hit, for, if the enemy is not held, not only may he attack and so disjoint our concentration, but he may shift his position so that, when we do strike out, our blow is ineffective.

When the enemy cannot be held, then, if concentration is to be effected, it must take place outside his reach. Such concentrations have sometimes to be resorted to during retirements, and the prevailing mistake made, as history will again and again show, is that, whilst the front is retiring, reserves are created piecemeal in rear and pushed into the battle, and destroyed in detail. If the enemy cannot be held, then the distance between him and the position of concentration selected must be sufficient to give ample time for the concentration of the forces required. In August 1914 the British Expeditionary Force was concentrated too far forward, seeing that the German right wing was virtually unopposed. At the end of this same month the French Sixth Army should, from the start, have been assembled at Paris, and not at Amiens. In March 1918, when the British Forces were driven back from the St. Quentin area, attempts were made to reinforce the defeated troops. It would have been sounder, I think, if the defeated forces had been left to fend for themselves, and that, in place, every available man had been concentrated well in rear, not to counter-attack, but to hold; for, from a moving base, to hit a moving enemy is almost as difficult as to attempt to shoot partridges from the window of a railway carriage. The ideal conditions in which concentration can accentuate offensive power is when a stable base of operations has been established, and the enemy has been forced to halt and so conform to the will of his adversary.

### 7. Mr. Lanchester's "N-Square Law"

A short time back I stated that superiority of numbers at the decisive point was not necessarily an application of the principle of concentration, since it is by means of weapons and not numbers of men that effect is obtained. If by superiority of weapon-power we can economize men, equally can we concentrate force. In his book, *Aircraft in Warfare: the Dawn of the Fourth Arm*, Mr. Lanchester, the eminent engineer, has, from a mathematical standpoint, examined the principle of concentration, and has contrasted, on a weapon basis, the conditions of ancient and modern warfare as follows:

Taking, first, the ancient conditions where man is opposed to man . . . there will be about equal numbers killed of the forces engaged; so that if 1,000 men meet 1,000 men, it is of little or no importance whether a "Blue" force of 1,000 men meet a "Red" force of 1,000 men in a single pitched battle, or whether the whole "Blue" force concentrates on 500 of the "Red" force, and, having annihilated them, turns its attention to the other half; there will, presuming the "Reds" stand their ground to the last, be half of the "Blue" force wiped out in the annihilation of the "Red" force in the first battle, and the second battle will start on terms of equality, i.e. 500 "Blue" against 500 "Red."

Now let us take the modern conditions. If, again, we assume equal individual fighting value, and the combatants otherwise (as to cover, etc.) on terms of equality, each man will in a given time score, on an average, a certain number of hits that are effective; consequently the number of men knocked out per unit time will be directly proportional to the numerical strength of the opposing force. Putting this in mathematical language, and employing symbol $b$ to represent the numerical strength of the "Blue" force and $r$ of the "Red," we have:

$$\frac{db}{dt} = -r \times c \; \ldots\ldots\ldots\ldots \; (1)$$

and

$$\frac{dr}{dt} = -b \times k \; \ldots\ldots\ldots\ldots \; (2)$$

in which $t$ is time and $c$ and $k$ are constants ($c = k$ if the fighting values of the individual units of the force are equal.)

A little later on Mr. Lanchester considers the efficiency of weapons, as follows:

Any difference in the efficiency of the weapons—for example, the accuracy or rapidity of rifle-fire—may be represented by a disparity in the constants $c$ and $k$ in equations (1) and (2). The case of the rifle or machine-gun is a simple example to take, inasmuch as comparative figures are easily obtained which may be said fairly to represent the fighting efficiency of the weapon. Now numerically equal forces will no longer be forces of equal strength; they will only be of equal strength if, when in combat, their losses result in no change in their numerical proportion. Thus, if a "Blue" force initially 500 strong, using a magazine rifle, attack a "Red" force of 1,000, armed with a single breech-loader, and after a certain time the "Blue" are found to have lost 100 against 200 loss by the "Red," the proportions of the forces will have suffered no change, and they may be regarded (due to the superiority of the "Blue" arms) as being of equal strength.

If the condition of equality is given by writing M as representing the efficiency or value of an individual unit of the "Blue" force, and N the same for the "Red," we have:

Rate of reduction of "Blue" force:

$$\frac{db}{dt} = -Nr \times \text{constant} \quad \ldots\ldots \quad (3)$$

and "Red"

$$\frac{dr}{dt} = -Mb \times \text{constant} \quad \ldots\ldots \quad (4)$$

And for the condition of equality:

$$\frac{db}{b\,dt} = \frac{dr}{r\,dt}$$

or

$$\frac{-Nr}{b} = \frac{-Mb}{r}$$

or

$$Nr^2 = Mb^2 \quad \ldots\ldots\ldots\ldots\ldots\ldots \quad (5)$$

In other words, the fighting strengths of the two forces are equal when the *square of the numerical strength multiplied by the fighting value of the individual units are equal.*

*The Outcome of the Investigation. The N-Square Law.* It is easy to show that this expression (5) may be interpreted more generally; the *fighting strength* of a force may be broadly defined as proportional to *the square of its numerical strength multiplied by the fighting value of its individual units. . . .*

*A Numerical Example.* As an example of the above, let us assume an army of 50,000 giving battle in turn to two armies of 40,000 and

30,000 respectively, equally well armed; then the strengths are equal, since $(50{,}000)^2 = (40{,}000)^2 + (30{,}000)^2$. If, on the other hand, the two smaller armies are given time to effect a junction, then the army of 50,000 will be overwhelmed, for the fighting strength of the opposing force, 70,000, is no longer equal, but is, in fact, nearly twice as great—namely, in the relation of 49 to 25. Superior *moral* or better tactics or a hundred and one other extraneous causes may intervene in practice to modify the issue, but this does not invalidate the mathematical statement.

*Example Involving Weapons of Different Effective Value.* Let us now take an example in which a difference in the fighting value of the unit is a factor. We will assume that, as a matter of experiment, one man employing a machine-gun can punish a target to the same extent in a given time as sixteen riflemen. What is the number of men armed with the machine-gun necessary to replace a battalion a thousand strong in the field? Taking the fighting value of a rifleman as unity, let $n$ = the number required. The fighting strength of the battalion is $(1{,}000)^2$, or:

$$n = \sqrt{\frac{1{,}000{,}000}{16}} = \frac{1{,}000}{4} = 250$$

one or quarter the number of the opposing force.

## 8. The Value of the "N-Square Law"

I have set down this long quotation with a purpose. Here is a noted scientist making use of mathematics to discover tactical truths. Mr. Lanchester fully realizes the weak points in his theory, so I must ask the reader, should he be inclined to criticize it, before doing so, to read his book. To myself, the main interest of the "n-square law" is that it enables us *in a certain extent* to arrive at the size of concentrations, and that, granted an equal *moral*, concentration of force is mainly to be sought for in weapon improvement.

For argument's sake, I will accept the statement that 250 machine-gunners possess the fighting power of 1,000 riflemen. How are we to proceed further? The answer is by thinking in the terms of the remaining two physical elements of war—movement and protection.

The machine-gun can only be fired from a stationary position. Suppose now that it be mounted on a cross-country tractor which will enable it to be moved and fired simultaneously, and that, consequently, its factor of efficiency is raised about three times—that is to say, from 16 to 49. Then the 250 men will be reduced to 143. Again, I will suppose that, by covering the tractor with bullet-proof armour plate, the factor of efficiency is increased from 49 to 400. Then we shall find that 50 men equipped with tanks have an equivalent fighting power to 1,000 riflemen. I will now suppose that these 50 men represent the crews of 10 machines, each machine being equipped with 4 machine-guns, and that these machines are ranged in battle against 1,000 riflemen. Turning to the war, we find that, even with the crude British tanks then used, on April 24, 1918, 7 Whippet machines, each holding three men and equipped with three machine-guns, with ease defeated 1,200 to 2,000 riflemen and infantry machine-gunners. We must, therefore, modify our factor of efficiency. I will assume that approximately half the 21 men are sufficient, then we obtain $\frac{1,000}{x} = 10$, therefore $x = 100$; therefore, in place of 400, the factor of efficiency is 10,000. With a tank moving at 20 miles an hour, in place of eight, it should be possible to reduce the figure 10 to 5. Then we get a factor of efficiency of 40,000, which brings us up to present-day possibilities.

What have we done? By improving weapons, movement, and protection, we have enabled 5 men, equipped with machine-guns, to equal the fighting power of 1,000 armed with the rifle. Will it be contended that an equivalent reduction in man-power could have been effected by some new process of moral training? No! The critic who bases everything on *moral* will not stick to his guns. He will, in place, abandon them in the face of the enemy, and assert that the above is not a fair example; that rifles are not the only weapons the tank will meet; that it will have to reckon with field-guns, and that the Great War proved that a single field-gun could knock out a whole company of tanks. It is, therefore, after all, a weapon which is going to beat the tank and not a man's heart. That the man who uses this gun must have the inclination to fire it and possess some skill in its manipulation goes without saying, and the higher his *moral* is the better will it be for the firer. Nevertheless, the fact remains true that it is weapons which do the work, and that it is this work which safeguards *moral*, which is always a doubtful quality, whilst the power of weapons is far more certain. In fact, he who possesses the superior weapon possesses the highest chance of victory.[1]

[1] In naval warfare the 99 per cent. weapon factor has long been realized. During the Great War there is only one recorded instance of a naval action in which a fleet of inferior weapon-power wilfully sought to engage one of superior force, namely Admiral Cradock's attack on von Spee's squadron at Coronel. Cradock possessed a slight superiority in speed, but a marked inferiority in gun-power. If ever *moral* had a chance of making good deficiency in weapon superiority, it was so in this action. Cradock's pluck is beyond criticism, nevertheless the Good Hope and Monmouth went to the bottom, not through an act of God, but through an act of mathematical certainty, and *von Spee did not lose one man.*

Mr. Lanchester's " n-square law " must be accepted as a most valuable idea possessed of a truth. We cannot slavishly follow it. My reason for having discussed it in detail is that, whilst formerly the application of the principle of concentration aimed at massing numbers of men, it should now aim at accentuating weapon-power. I will therefore end this section with another quotation, also big with truth. In *Sartor Resartus* Thomas Carlyle writes:

> Such I hold to be the genuine use of gun powder; that it makes all men alike tall. Nay, if thou be cooler, cleverer than I, if thou have more *mind*, though all but no *body* whatever, then canst thou kill me first, and art the taller. Hereby at last is the Goliath powerless, and the David resistless; savage animalism is nothing, inventive spiritualism is all.

Thus mind triumphs over matter, and the body is its tool.

### 9. The Principle of Surprise

Concentration of force is first an act of will, and, secondly, a massing of means, and I have just shown the enormous importance of means in the application of this principle. To-day one modern cruiser could sink the whole of Napoleon's fleet at Trafalgar in an hour or two with no loss or inconvenience to itself, and, though we cannot hope to attain such weapon superiority over an enemy, we should realize that it is through concentration of thought this superiority is attained, and that the nation which does attain this superiority, irrespective of its man-power, can proportionately increase its force on the battlefield.

It must not, however, be overlooked that weapons, however powerful they may be, are useless unless the will can direct them. This direction depends on knowledge, on skill, and on sentiment, consequently these three qualities must exist in an army if the full power of the weapons is to be developed. Sentiment must be such that knowledge and skill can operate. The ultimate expenditure of force, as I have shown, depends on the determination of the soldier. If this determination is reduced to zero, then his power to wield weapons with skill and knowledge becomes negligible; in fact, his power to expend his force economically is reduced to vanishing-point. The control of all the conditions of war which so influence a man's will that it loses its determination to exert pressure and resistance is the province of the principle of surprise.

In war, as I have explained, force can seldom, if ever, be

directed in a straight line, and that, consequently, from the physical aspect of concentration, the side which can exert superior pressure against inferior resistance is the side which is more likely to succeed. In the moral aspect, however, if resistance be deprived of its endurance by the application of surprise, then frequently a physically inferior force will be able to overthrow a physically superior one, because its unexpected action will have created a line of moral least resistance.

The subject of surprise is an immense one, and one which influences all forms and modes of war. It is one which is nearly always lost sight of during peace-time, because danger and fear are more often than not abstract quantities; but in war-time they manifest, and with them manifests surprise—the demoralizing principle. Clausewitz must have recognized this when he wrote: "Has not then the French Revolution fallen upon us in the midst of the fancied security of our old system of war, and driven us from Chalons to Moscow? And did not Frederick the Great in like manner surprise the Austrians reposing in their ancient habits of war, and make their monarchy tremble? Woe to the Cabinet which, with a shilly-shally policy, and a routine-ridden military system, meets with an adversary who, like the rude element, knows no other law than that of his intrinsic force. Every deficiency in energy and exertion is then a weight in the scales in favour of the enemy; it is not so easy then to change from the fencing posture into that of an athlete, and a slight blow is often sufficient to knock down the whole."[1]

In war surprise is omnipresent; wherever man is there lurks the possibility of surprise, yet it is intangible and all but omnipotent. From this it will be understood that in the few pages at my disposal I cannot do more than touch the fringe of this all-pervading principle, and because of this I must urge the student to do more than merely read my words. He, if he wishes to understand war, must examine the nature of surprise in its thousand and one forms as it pursues its restless course through history.

Without surprise in some form or another it is not possible to maintain the law of economy of force. Even if I have one hundred men and am opposed by one man, I must apply this principle, for if, in killing or capturing this one man, I lose two or three men, when, in the circumstances, by applying surprise I might have sustained no loss at all, then I shall have violated economy of force.

Surprise should be regarded as the soul of every operation. It is the secret of victory and the key to success. It originates

[1] *On War*, vol. iii., pp. 229, 230.

in the mind of man and accentuates the power of his will; it is the weapon of intelligence, this harnessing of fear. As direction springs from the mind, so does surprise spring from the sentiments. It has power over *moral*, and can raise or depress it instantaneously, accordingly as it is used by us or against us. It can destroy *moral* as rapidly as with a pin I can destroy a soap-bubble; and, above all, it is a double-edged tool, and an exceedingly dangerous one in clumsy hands, for few disasters are greater than the surprisal of a would-be surpriser. Panic is never more latent than when one side imagines it has victory by the throat.

### 10. The Means of Surprise

The object of surprise is to attack the will of the enemy by accentuating fear, for, if a man is reduced to such a state of fear that he can do nothing save think of protection, he is at our mercy, for his moral endurance has ceased to dominate him. A man whose mind is dominated by fear is a man in panic, consequently the ultimate end of surprise is to reduce our enemy to a condition of panic in which his moral is totally replaced by his instinct of self-preservation in its most irrational form.

The conditions of surprise are innumerable, but the means may be classed in three great categories, namely the mental, moral, and physical. Thus:

(i.) Surprise effected by superior direction.
(ii.) Surprise effected by superior determination.
(iii.) Surprise effected by superior mobility.

The first is based on distribution of force, and is expressed through concentration of force; the second on moral endurance, and is expressed through power to demoralize; and the third is based on security, and is expressed through offensive action.

The means of surprise are those which spring from the ability of the general, the courage of his men, and the perfection of their weapons.

Though none of these are constant, for their values are always changing, by far the most difficult to gauge is the first—it is the dark horse of the battlefield. Mental ability is not so much a natural gift, save in the case of very few, as the product of scientific study—a close reasoning out of the values of conditions and an intelligent application of the principles of war. Again, mental ability does not so much consist in inventing superior weapons, means of movement and protection, as in combining the existing means according to their true values. What are their values? It is here that mistakes are being persistently made. In 1870, because the *mitrailleuse* was mounted on a gun-carriage, the French employed it like a field-gun; in 1914, because the Vickers machine-gun fired .303 bullets, we employed it like a rifle. The machine-gun is neither a field-gun nor a rifle, for it is a machine-gun, and very different from both these other weapons. It has tactics of its own, and because, in 1914, all parties were hallucinated by rifles and field-guns, its value remained hidden, and the discovery of its value proved one of the most costly surprises of the war.

In 1914 all parties were surprised by fire-power. In a few weeks the tactics of forty years were divested of all semblance of utility; they might just as well have never been written; in fact, in many cases they proved disastrous deterrents to common sense action, for thousands of lives were lost in trying to apply them. The war opening with this colossal surprise, all sides were smitten down by a paralytic stroke, and the war grew rigidly static. From November 1914 on to the spring of 1918 all sides searched every nook and cranny of the art of war for the secret of surprise. Ultimately, from March 1918 onwards, one surprise followed another, and the object of the Allies—the defeat of the German armies—was gained through a series of surprises which palsied the will of the German nation and caused the foundations of the German armies to crumble and give way.

### II. Tactical and Strategical Surprise

An appreciation of the true values of the physical means of war is, therefore, as we see, the foundations of surprise, which I will now consider in its tactical and strategical forms, concerning which Marshal Foch says:

> "Whatever a thing may be," writes Xenophon, "be it pleasant or terrible, the less it has been foreseen the more it pleases or frightens. This is seen nowhere better than in war, where *surprise* strikes with terror even those who are much the stronger party."
>
> . . . . . . . . . . . .
>
> The means of breaking the enemy's spirit, of proving to him that his cause is lost, is, then, surprise in the widest sense of that word.
>
> Surprise bringing into the struggle something "unexpected and terrible" (Xenophon); "everything unexpected is of great effect" (Frederick). Surprise depriving the enemy of the possibility of reflection and therefore of discussion.
>
> Here we have a novel instrument, and one capable of destructive power beyond all knowledge. However, one cannot obtain this at will; setting an ambush, attacking in reverse, are possible in a small

war, but impracticable in a great one; it is necessary, therefore, to resort in case of great wars to bringing forth a danger which the enemy shall not have the time to parry or which he shall not be able to parry sufficiently. A destructive force must be able to appear which should be known, or seem to, the enemy to be superior to his own; to this end, forces and thereby undisputable efforts must be concentrated on a point where the enemy is not able to *parry* instantly—that is, to answer by deploying an equal number of forces at the same time. Such will be our conclusion.[1]

It is quite true that an able commander will, whenever he can, attempt to bring forth a danger which the enemy is unable to parry; it is also true that we can seldom hope to ambush an army; but that in great wars armies cannot be attached in reverse is not borne out by history. At Jena, Napoleon attacked the Prussians in reverse, and in 1914, had von Moltke shown normal aptitude, the whole of the five French armies would have been attacked in reverse and, in all probability, have been swept into Switzerland. The reverse, or rear, attack is, in fact, the supreme surprise operation not only of small wars, but of great. I will now examine this form of surprise.

The military will of an army is centred in its command—its brain. This will is based on the national will behind it, and is protected by the will of the soldiers engaged. We are confronted here by a very interesting problem, which I will explain diagrammatically.

| Combatants |
|---|
| Reserves |
| Command |
| Government |
| Nation |

The diagram may be looked upon as a tower, the Nation is its ground floor, and the Government, Command, Reserves, and Combatants its four storeys. If the tower is to be demolished, the speediest way is to blow up its foundations. If this is impossible, then to blow up its first storey; if impossible, its second; if impossible, its third; and, if the fourth storey can only be attacked, then the process of demolition becomes very slow. In war the last method has been the normal method. I

[1] *The Principles of War*, pp. 291–92.

believe that the power of aircraft will render it less normal, but I will here exclude this possibility, and only consider the influence of demoralization on the three top storeys.

The combatants are in movement, they are pushing forward or being pushed back. They are faced by the reality of battle and know what is happening. The kaleidoscope of events is changing so rapidly that time is normally insufficient for their thoughts to concentrate for long on fear.

From those actually engaged, turn to the reserves. They are halted. They are surrounded by images and not by actualities. They know that a battle is being waged in front of them, but they are out of touch with its reality. Time for brooding is ample; bad news travels swiftly, and fear is contagious. Curious as it may seem, though they are not fighting, they are frequently more susceptible to demoralization than those engaged. The uncertainty of the unknown is sapping their *moral*. They are like men looking into a convex mirror, the further back they withdraw their heads the more distorted becomes the reflection, until ultimately nothing is seen clearly. What does this teach us? It teaches us that the rear demoralizes the front; that to surprise the front we must attack the rear. First the rear of the front, secondly the rear of the reserves, thirdly the rear of the command, and so on back to the initial will of the people who desire victory and dread defeat.

As physical weapons hit fronts, so do moral weapons hit backs, and the most potent of moral "weapons" is surprise. The interplay between these two weapons forms the backbone of the attack. In the normal physical attack the decisive point is a physically weak point—a point which can be easily attacked and which it is difficult for the enemy to protect or to reinforce by means of his reserves. As the lack of reserves is the normal condition which constitutes physical weakness, physically weak points are generally those which are distant from the reserves. In the moral attack—that is, an attack in which brute force is replaced by surprise—this condition does not necessarily hold good, for frequently the morally weak point is one which is closely supported by reserves.

The reason for this is that, if the enemy's front can be rapidly disorganised by a surprise-attack, its shattered fragments, like the jagged pieces of some immense shell, will strike the reserves and morally tear them to pieces.

It may be thought that there is some "catch" in the logic of this argument, that this is not really so; but history will prove that it is so. In the files of the Grecian phalanx the bravest man was in front and the next bravest in rear; in the Roman legion the *triarii*—the veteran troops—were in reserve; in the army of Napoleon the Old Guard was in reserve. If in war we are faced by an enemy who places his best troops in front and holds his worst in reserve, the moral point of attack will be opposite his reserves; they will constitute a human explosive which at any moment may detonate and blow him to pieces. If, however, he holds his best troops in reserve, we must be on our guard where we attack him. If these troops be veterans, we must be doubly on our guard, for they know what the reality of war is; if they be young and inexperienced, we may accept risks and act with audacity. Here, then, is our ultimate conclusion; the decisive point is the normally most sensitive point and not the numerically weakest point, and the weapon of the moral attack is surprise.

The point, I think, that Marshal Foch overlooked when he wrote his book was the extreme importance of strategical surprise, which renders tactical surprise on the grand scale possible. Clausewitz was much more certain on this point. He says: "In tactics, a surprise seldom rises to the level of a great victory, while in strategy it often finishes the war at one stroke"[1]; which is very true; and it was such a surprise which very nearly took effect in 1914.

Whilst tactically we attempt to hit at moral objectives, strategically we try to manœuvre towards and into "moral" spaces, and normally these spaces are those which include the communications of the enemy's forces to be attacked. Here we are confronted, not by the rear attack, but by the rear manœuvre which either culminates in battle or in a change of communications. In August 1914 the French in Lorraine tried to strike at the German communications north of Strasburg, and the Germans, meanwhile, by moving through Belgium, struck at the French communications between Lorraine and Paris. The first result was the change of the British communications from Havre to St. Nazaire, and the second the battle of the Marne. To the Allies the immediate object of this battle was to secure their communications, which had been surprised.

The chief means of strategical surprise are:

(i.) Simplicity of movement.
(ii.) Secrecy of movement.
(iii.) Speed of movement.

Morally weak spaces can be created by many means, such as misleading the enemy, pandering to his stupidity, leaving unprotected enticing lines of advance, moving by unexpected lines of

[1] *On War*, vol. ii., p. 144.

approach, and threatening vital points without any intention of attacking them. When we study the campaigns of Napoleon we find innumerable cases of the strategical surprise. His Italian campaigns are full of such cases, and startling examples may be discovered in the Marengo and Jena campaigns. Hannibal was another master of the strategical surprise, and so, in a lesser degree, was Marlborough.

### 12. The Influence of Scientific Weapons on Surprise

The Great War of 1914–18 was remarkable in many ways—the size of the forces contending, the lack of able leaders, the stupendous fire-power utilized, the development of aircraft, and the general utilization of the petrol engine; but, beyond all these, scientific invention surpassed anything dreamt of in former wars. The Franco-Prussian War, the South African War, and the Russo-Japanese War were won with the weapons of the mobilization stores. In the last-mentioned war a few minor inventions were introduced, and the power of existing weapons improved.

In the Great War, partly due to its length, but mainly because it was fought by nations possessing immense scientific knowledge, invention followed invention, and many existing weapons were improved beyond recognition. So much so was this the case that, had the war lasted another two years, the equipment of 1914 would have been completely replaced, and an entirely new epoch of war would have opened, based mainly on the aeroplane and the cross-country tractor, with gas as the superior weapon.

During the war each new invention ushered in the possibility of surprise, but this possibility was seldom grasped, because no method existed whereby the soldier could discover, save by the slow process of trial and error, the tactical value of any new weapon. The principle of surprise was violated again and again through sheer ignorance. The means existed, but ability to understand these means was lacking. The higher command of all armies never grasped their scientific limitations, and for the following reasons: because they had been brought up in a school of war the doctrines and methods of which bore little resemblance to reality; because seniority carried with it a fictitious omniscience; and because totally ignorant men would again and again wave aside, with a gesture of pitiful sorrow, the opinions of the highest experts. In spite of this lack of power to grasp the values of things new, inventions played such a preponderating part during 1918 that two totally false points of view were established. The first was that the higher command had shown consummate skill, and the second that in the face of new inventions skill is next to useless. If these erroneous opinions are allowed to persist, then one thing is certain, namely, that the next war will produce a series of surprisals unprecedented in history. All sides will surprise each other with their eyes wide open, and the greatest surprisals will be effected when they are least intended. For the historian it will be a war of much interest and perplexity—a war of flukes.

### 13. The Influence of Tactical Organization on Surprise

In my opinion both these points of view can be proved false—the first in that, had the higher commanders really shown ability in the use of inventions, they could not, immediately peace was declared, have reverted so rapidly to the 1914 organization and equipment; the second in that it is manifestly wrong to place present-day inventions in a separate category to those which preceded them. If in the past skill has been able to utilize weapons, in the future it will again be able to do so. There is, however, this possible difference: whilst in the past the mobilization equipment of civilized armies was generally known, in the future certain very important items may be unknown, and only become known on the battlefield. It is conceivable that a discovery may be made by one nation during peace-time which is so overwhelmingly powerful that no enemy unequipped with it could hope to conquer. If this be the case, then it points to the vital necessity of foreseeing the future under all possible shapes and forms, of liberally using hypotheses of victory, and by every means possible proving them false or true. Excepting this category, the bulk of inventions will be known; consequently, as heretofore, tactical skill in their use will play an important part. In the past this skill has manifested itself during war, because tactical organization was extremely simple. Then there were three simple arms—infantry, cavalry, and artillery—and three simple weapons—rifles, swords (or lances), and field-guns.

Tactical organization was based, therefore, on the following plan: whilst the guns protected the infantry, the infantry attacked the enemy's infantry, and when the enemy was demoralized the cavalry charged home and annihilated him. To-day we maintain this organization, but it is visibly out of date. Where to-day do the new inventions of 1918 fit in? They do not fit in, so they are appended to it. To take a simile. The Saxon with his battle-axe is equipped with the bow, equipped with the cross-bow, equipped with the arquebus, equipped with the musket, equipped with the rifle, and finally with the Lewis-gun. He steps on to the battlefield a veritable museum. He cannot use his battle-axe because of the other "appendices." One moment he wants to use his axe and he falls over his rifle, the next moment he wants to use his Lewis-gun and he trips over his bow. He cannot combine their powers.

What I wish to point out here is that it is not the enemy who is surprising him; he is surprising himself, because he is not organized to do anything else. He is enmeshed in surprises, and is astonished each time he attempts to make use of one of his appended weapons.

The lesson to be learnt is that tactical organization is one of the main props of the principle of surprise. I have shown that the principle of direction is derived from the elements of the mind, and that this principle finds expression in the determination of the will of the commander. The principle of surprise accentuates or destroys determination. For the will to attain its end, its means of expression—the moral elements of war—must be organized—that is, set together in such a manner that the highest economy of force can be effected through the harmony of their joint values. If this harmony does not exist, how can originality of thought, which leads to surprise, accentuate the will? It cannot; it can only confuse it. A man who speaks ten foreign languages organizes his brain to work economically in the ten countries in which these languages are spoken. What does the modern soldier do? He learns one language and puts nine dictionaries into his haversack. He then steps on to the battlefield and, if the language he knows is not understood, he opens his dictionaries and misuses words, with surprising results—says *cochon* when he means *cocher*, and, if he is so fortunate as to beat his enemy, he presents himself with a first-class interpretership.

I have accentuated this relationship between organization and surprise because, if we examine past history, we shall find it has played such an obscurely decisive part. We equip ourselves with new weapons, but we fail to discover their values or the relationship between their respective values. We invent tactics on suppositions, and then we organize our forces to fit traditions, barrack-rooms, parade grounds, and certain round sums of money. Worse still, if we succeed in one war we imagine that to copy our success is the panacea against future defeat. What we must do, if we wish to prepare ourselves to apply the principle of surprise and to secure ourselves against its application, is to cease thinking in terms of infantry, cavalry, and guns, and to think in terms of the elements of war—to take each weapon and extract its values, to take all these values and extract their relationships, and then finally, having evalued our future enemy, to organize these relationships into a tactical whole—that is, an UNIT—which possesses the maximum offensive power, protective power, and moving power, because these powers when *combined* are *one power*—tactical power—and not three powers, or three collections of power, hung on the skeleton of a military organization which was found efficient in the days of von Moltke, or Wellington, or Sennacherib.

### 14. The Sword and Shield of Surprise

The main causes of surprise are lack of foresight, loss in sensing the reality of war, lack in appreciating tactical values, and, above all, the strangling grip of tradition which is ever choking our intelligence.

To copy is not to originate, and originality of thought is the mental co-efficient of the principle of surprise, and, when determination to win is accentuated by this principle, *frequently an objective can be created by one side which is totally unrealized by the other.* Such a creation is what I call tactical forethought—seeing an action before it is fought. Foresight is the fruit of the scientific method, and it must not be confounded with imagination. Imagination presents to us a possibility, reason analyses it and stamps it with a value; these two are the parents of foresight, which is nothing more than mentally standing on tiptoe.

This is the main application of this principle, whether it be used protectively or offensively. Whatever we will to do, we must foresee what is most likely to happen. There is always one supremely right thing which we should do, but it is usually hidden. We must discover it, and we shall never discover it as long as we remain slaves to the past and pour out our oblations before archaic idols.

The scientific method is the surest means of preparing for or against surprisals, as it enables us to arrive at true values. It is method which we require, a method based on judgment and not on dogma. To lay down a method of procedure is the matter of a moment, but to work out the results of such a method is the task of years, and to establish a common doctrine on these results may take several generations. If this be true, then we must make absolutely certain of existing values before we attempt to forecast future influences. If we do so, if we establish a scientific base of operation, then the results of our method will unfold themselves systematically, one result pointing the way to the next, and each rectifying the method itself.

In war it takes time to gain superiority in anything, and time is nearly always at a discount; consequently we find, although minor surprisals may be accomplished by seizing opportunity, the possibility of effecting major surprises depends mainly on the forecasts and preparations which we have made during days of peace. The surest foundation of eventually being surprised is to suppose that the next war will be like the last war, and that consequently old means will accomplish new ends. The general who slavishly copies former battle tactics is more often than not surprised with his eyes wide open. He sees things coming, but, blinded by prejudice and hallucinated by tradition, he does not perceive their consequences, because he cannot appreciate their values. Even when routed again and again he cannot trace cause and effect; he attributes his defeat to some unconnected incident, attempts to copy it, and is defeated again, and yet again.

On the battlefield itself a general is frequently surprised by his own stupidity, his lack of being able to appreciate conditions or apply to them the principles of war. This stupidity sometimes takes the acute form of completely misunderstanding human endurance. Not realizing what they can do, the troops are ordered to do something which they cannot do, and the result is chaos and loss of life. It is indeed a curious contemplation that, whilst a progressive and warlike nation will go to infinite trouble to drill its army to perfection and spare no cost in its equipment, no army has hitherto *scientifically* prepared itself to meet or to effect surprise. With a few elementary rules and a pinch of military jargon any intelligent man can become what is called a "strategist" or a "tactician." In the last war, like every other war before it, every other man considered himself a military authority, and, in fact, he was one, and will continue to remain one as long as the alchemical epoch of war endures. In no other science could such an outlook exist. In biology, chemistry, mathematics, mechanics, and astronomy the expert stands apart from the amateur and the ignorant, and why? Because he has accumulated knowledge scientifically, and they have only gleaned bits here and there. As long as we remain amateurs we shall be surprised, sometimes by the substance of the enemy, but more often by the shadows of our ignorance.

### 15. The Principle of Offensive Action

I now come to the third principle of pressure—the principle of offensive action—a principle which has been so thoroughly misunderstood since the Prussian System of war began to dominate military thought; for, according to this system, the deciding force is not the intelligence of the general, but the brute force of his men. The result of this system has been that during the last seventy years, even more so than the years which preceded its acceptance, warfare has become throughly brutalized.

The object of the offensive is not to kill, wound, and capture, but to establish a condition which will permit of policy taking effect. But, as Clausewitz says: "Activity in war is movement in a resistant medium,"[1] consequently this condition cannot be established until resistance is overcome, and the overcoming of resistance demands destruction of hostile force, for "the essence of the attack is movement,"[2] and until this resistance is removed freedom of movement is not possible, and unless movement is free the will of the general must remain shackled. Further, Clausewitz writes: "Activity in war is never directed solely against matter; it is always at the same time directed against the intelligent force which gives life to the matter, and to separate the two from each other is impossible."[3] In the conception of victory he finds "three elements," namely:

(i.) The greater loss of the enemy in physical power.
(ii.) The moral power.
(iii.) His open avowal of this by the relinquishment of his intentions.[4]

The third element is, as we see, loss in mental power.

The first of these losses is accomplished by means of physical force, and the second by that of the moral attack. The question now arises, Which of these two means is the most economical? For the least economical will violate the law of economy of force. From what I have said about surprise, the answer undoubtedly is that in expenditure of force the moral attack is undoubtedly more economical than the physical attack, therefore the true object of the attack is to strike at the enemy's determination to continue to resist, for when his determination is broken his direction ceases to control and he is compelled to relinquish his intention. The aim of the principle of offensive action is, therefore, to compel the enemy to accept our will with the least expenditure of force. The offensive is, consequently, not merely a brutal act, but largely an intelligent act.

### 16. The Direction of the Offensive

The offensive is a mental, moral, and physical act. The "will to win" is the driving force, the "power to endure" the staying force, and the "ability to kill and to terrify" the deciding

[1] *On War*, vol. i., p. 79. [2] *Ibid.*, vol. ii., p. 9.
[3] *Ibid.*, vol. i., p. 101. [4] *Ibid.*, vol. i., p. 250.

force. In other words, the offensive is the application of will-power by moral and physical means. If any one of these three factors be deficient, the remaining two are useless. The moral is not to the physical as three to one, neither is the physical to the mechanical as one to three, for each in itself is useless without the other two, and to juggle in the proportions of essential qualities is of little help.

The object of the offensive is to destroy the enemy's strength, which is centred in his will to command, and which finds expression in the organization of his forces and endurance in the *moral* of his men. Organization enables the will to express its intention rapidly and without friction, to concentrate the means it uses, and to amplify their power. Organization is, in fact, the medium of command ; further, it endows *moral* with solidarity by rendering unity of action possible.

To apply the principle of offensive action is to break down this unity by disorganizing the enemy, which may be accomplished by attacking the physical or moral foundations of his army. In the first case, the destruction of order is brought about by the application of brute force, and, in the second, by fear and terror, leading to panic. The second means are incomparably more economical than the first. In the first case, if we kill a man the dead man cannot in his turn kill one of his fellows, but even when dead he can demoralize him by bearing witness to the power possessed by the enemy to inflict death. If the dead be removed from the living, this demoralizing influence to a great extent ceases. In the second case, if we terrify a man he becomes a mobile demoralizing agent, and if we terrify a number of men the probability is that they will seek relief from terror by quitting the field of action, and, as their line of retirement will normally lead them towards the troops in rear, a panic may result ; for, as I explained when I examined the principle of surprise, the *moral* of the reserves is frequently in an unstable condition.

In the past, on account of the restricted range of weapons, it was only possible to strike at the rear of an army by penetrating its front or by manœuvring round its flanks, for all offensive action took place on a plane surface. With the increase of gun-range, by degrees it became possible simultaneously to attack the rear and front of an army from a static position. To-day the aeroplane has rendered this position dynamic, and has given such range to the rear attack that it is possible to picture the whole of a reserve army being annihilated whilst the forces in front of it are not even engaged. I do not intend to pursue this argument, and I only mention it in order to accentuate the fact that, whilst the application of the principle of offensive action was limited by the conditions inherent in two-dimensional warfare, to-day the possibility of adding a third dimension, though it has in no way altered the principle, has vastly extended its application. As the powers of aircraft grow the whole of our military organization will have to be recast, for in a two-dimensional organization it will be next to impossible for the will of a commander to find expression if he is opposed by a third-dimensional weapon. The solution to this problem does not concern us here, but it may not be out of place to mention that it will not be discovered by appending aircraft to land forces—infantry, artillery, etc.—but, in place, by examining all existing means—aeroplanes, infantry, artillery, etc.—from the point of view of the elements of war, extracting their values, discovering their relationships, and then creating an organization through which the will of the commander can find its highest expression in their use. What the commander of the future must aim at is the accomplishment of the offensive through mental paralysis as well as through physical destruction. He must understand the relationship between the power of weapons and the endurance of *moral*, and organize his forces on this relationship in place of on the various types of men who manipulate weapons.

### 17. The Relationship of the Offensive and Objective

Without mental activity we can accomplish nothing, and from mental activity arises physical action which, when directed against opposition with the intention of overcoming it, is in war called the offensive. In order to conform to the law of economy of force it may be accepted as an axiom that offensive power can never be too strong. Strength does not lie, however, in offensive action alone, but rather in protected offensive action—that is, action springing from a sound and secure foundation. If the attacker cannot be attacked, complete freedom of action is at his command, and, though this ideal can seldom be reached, the nearer we approach it the more powerful will become our offensive. The principle which governs the relationship between offensive action and security is that of distribution of force. The correct application of this principle enables us first to distribute force so that a secure base of operations is established from which offensive action can operate, and, secondly, it enables us to protect this offensive action itself as it runs its course.

Once this distribution is made, the success of offensive action is governed to a great extent by the choice of objective and by the conditions which hedge it round, conditions which will assist or resist the attacker. A general will seldom win without attacking, and he will seldom attack correctly unless he has chosen his objective with reference to the principles of war, and unless his attack is based on these principles. Imagination is a great detective, but imagination which is not based upon the sound foundations of reason is at best but a capricious leader. Even genius itself, unless it be stiffened by powerful weapons, a high *moral*, discipline, and training, can only be likened to a marksman armed with a blunderbuss—ability wasted through insufficiency of means. Conversely, an efficient army led by an antiquated soldier may be compared to a machine-gun in the hands of an arbalister. Will the objective that we have selected enable us to apply the principle of offensive action? If it will not, then the objective must be discarded, for the offensive in war is the surest road to success. If it will, then in which direction should the offensive be made? The answer to this question does not only depend on conditions, which should be looked upon as the correctors of all movements, but on power to apply surprise. An objective which cannot be attacked in daylight may frequently be attacked and surprised under cover of darkness. Again, the most apparent line of approach is not necessarily the line of least resistance.

### 18. The Anatomy of Offensive Action

In chapter vii. I stated that in battle confidence depended on certain psychological factors. These I will now amplify and examine more fully, for the psychological base of the offensive is the determination of the attackers. As the fighter is urged forward by his will, the attainment of the objective must demand of his will not more than his will can accomplish. The first characteristic of what I will call " the compound of secure movement " is, therefore, *limitation of the objective*. The objective must, so to speak, lie within a circle the radius of which is the maximum will-power of the man. It must also lie within another circle the radius of which is *moral ability to endure*, and yet a third—*physical ability to accomplish*. These two form the second and third characteristics in " the compound of secure movement." From these characteristics we can extract a further series. Will is charged like an accumulator by encouragement, which is fostered by a feeling of superiority begotten by continuity of policy (maintenance of the object), depth of formation, and superiority of weapons. The fourth characteristic may, therefore, be denoted as *stimulus of success.* This enthusiasm, which is always of a volatile nature, requires protection, and not only protection, but an uninterrupted flow;. the fifth and sixth characteristics are, therefore, *security of movement* and *continuity of action.*

As human nature, on account of the exhaustion and reaction which always follows strenuous work, demands at least temporary *immunity from danger,* this immunity becomes the seventh characteristic. When once a body of men has become exhausted, offensive action must be fortified—that is, it must be continued by fresh troops. The mere act of seeing fresh troops advancing beyond them, and so automatically protecting them, will, by securing their bodies from danger, refresh their minds. This brings us to our eighth characteristic—*the progressive base of operations.*

Whether an offensive be carried out over open field land or against a strongly fortified position, its foundations are to be sought in the base of operations from which the attack is launched. In the past this base has been considered as the original starting-line, and, if battles can be won in a single onslaught, this assumption is correct. As this is seldom the case, and as battles are normally won by relays of attacks, each echelon must start from a secure base; consequently there must be a base of operations to each objective, requiring a fresh echelon of troops. Each echelon of troops must be sufficiently self-contained, not only to be in a position to capture an objective, but to hold it once it is captured, and so form a base of operations for the echelon following it. Further, each echelon must be protected by the one in front of it as well as by those behind it and on its flanks, and, as the first echelon cannot be so protected, and the last is often similarly situated, it is essential that the leading troops and those which will form the ultimate battle-front should belong to a *corps d'élite,* the former setting the pace, the latter clinching the argument.

## 19. The Strategical Offensive

There are three main categories of offensive action: the ethical (moral), economic, and military attacks. All have frequently been used, and especially so during the period of mediæval warfare. Then we find the Pope using interdict and excommunication as weapons, and captured towns being handed over to the soldiery, not only to satisfy their lust and greed, but to terrorize whole districts. In recent years the moral attack on the nation itself has fallen into abeyance, and rules have been devised to restrict economic injury, but in the last great war all these restrictions were cast aside and all categories became active.

We may, consequently, expect that this fullness of war will continue, and that military attacks will be reduced in importance and take their place alongside the remaining categories. Be this as it may, I do not intend in this chapter to go outside military action.

There are two great classes of the offensive; the first is based on secure movement, and the second on secure offensive power. The first, in the main, belongs to the strategical offensive, and the second to the tactical. The one may form the base of the other. Thus a tactical offensive may be delivered in order to hold an enemy or draw him out of an area, so that liberty of movement may be gained for strategical manœuvre; or else a manœuvre may be made in order to threaten an enemy and force him to support the point threatened by withdrawing troops from an area in which it is intended to deliver a tactical blow. When these two types of offensive operations are attempted simultaneously they should be most closely related, one influencing and assisting the other, like the right and left hand punches of a boxer. Frequently a campaign is opened by a strategical offensive which culminates in a tactical operation. When this is the case, the object is either to draw the enemy into an area in which more profitable tactical action may be sought, or to draw an enemy away from his communications and then force him to fight for their security.

It is a mistake of the first order to believe that the seizing of the tactical initiative is of necessity the maintenance of the principle of offensive action. Though in many circumstances this is so, the initiative does not necessarily depend on attacking, but quite as much on manœuvring, until a situation is created where in a profitable attack may be driven home. To seize the initiative at the beginning of a campaign, unless the enemy be considerably weaker than oneself, often means that, before the campaign is a few weeks old, the initiative will pass to the enemy, because the conditions which surround the initial stages of a campaign are normally most difficult to gauge. If the initiative has to be seized, as was the case with Germany in 1914, then the only safe method of procedure is to maintain a large reserve in hand, so that initial mistakes may be rectified. The power to maintain initiative depends in most cases on the holding of strong reserves in hand rather than in attempting overwhelming attacks. I have already dealt with this subject under the principle of concentration.

## 20. The Tactical Offensive

The tactical offensive may roughly be divided into two classes. The first is governed by liberty of movement, and the second by restriction of movement.

Attacks based on liberty of movement may be divided into direct attacks and delaying attacks, and normally these are combined. Thus, in a direct attack our object is to march on the enemy and defeat him wherever he is; while in a delaying attack we march on him to halt him, to restrict his movement, so that the direct attack may take place on a selected battlefield. Generally speaking, the principle of offensive action is applied by first delaying the enemy—that is, restricting his power of movement—and, secondly, by pinning him down or fixing him—that is, by forcing him to assume a protective attitude—and, thirdly, by attacking him in superior force at a physically or morally weak point.

In order to restrict the advance of the enemy in a certain direction we must either directly bar his progress or we must force him to halt or change direction by threatening one or both of his flanks, or, better still, his rear. Or, again, if the hostile army in question is operating with another army, by attacking this other army we may force the first to withdraw. It will be seen, even from these few remarks, that it is not possible to lay down definite rules of attack, because it would be the exception for circumstances to admit of rules being applied. Each campaign and each battle requires a method of its own, but this method is governed by the principle of offensive action, which requires that the attack be delivered from a secure base, and be directed against a weak point, and protected until this point is pierced or shattered. The unlimited offensive—that is, an *offensive à outrance*—has nothing whatever to do with scientific warfare. Sometimes it may succeed by overwhelming a terrified antagonist, but if the enemy is alert and courageous it nearly always fails through premature exhaustion. Seldom will it be possible to march straight towards the enemy and defeat him, consequently many acts may have to be played before the curtain of victory is finally rung down. In scientific warfare each act must constitute a distinct and profitable step towards the transformation scene of peace. If this be not done, then an infringement of the law of economy of force will take place. This must be guarded against, for each blow must form a definite link in an offensive chain of blows, in which moves, as in chess, are seen ahead. Only when the enemy's endurance is exhausted, when his organization is shattered and his *moral* is verging on freezing-point, is an *offensive à outrance* justified in the form of a relentless pursuit, which is not so much an act of scientific warfare as of pure brute force—of courage, audacity, and endurance.

Attacks based on restricted movement are, more frequently than not, parallel actions of attrition. Here, again, the principle of offensive action can be incorrectly applied. We violate this principle and the principle of endurance if, possessing more men than brains, our object is simply to kill as many of the enemy as we can, regardless of cost. A private soldier thinks in terms of killing men, but a general should think in terms of destroying or paralysing armies. "Push of pikes" is a simple game compared to defeating an army, which requires an acuter intellect than that of a lusty halbardier.

In the last great war so many battles of attrition were fought that it is, I think, worth while examining this form of attack. A study of Napoleon's tactics shows clearly that when he was compelled to deliver a frontal attack, before attempting to break his enemy's front he first drew in the hostile reserves and disorganized them, his aim being to avoid any risk of being taken at a disadvantage. Once this was accomplished (and he also aimed at it in his battles of envelopment) no further opposition was to be expected, consequently a pursuit could be carried out, a pursuit being, more often than not, initiated by troops disorganized by victory against troops disorganized by defeat.

To turn to two examples in the Great War. Before the third battle of Ypres had begun, we had, through offensive action, forced the enemy to draw largely on his reserves. This, judged by the Napoleonic standard, was correct. Where we failed was that, once we had drawn these reserves in, we had no Old Guard at hand to smash them. At the battle of Cambrai we struck with our Old Guard (tanks) before the German reserves were on the battlefield. It was a blow in the air, and the result was that we crashed through the enemy's front, and then, when the enemy's organized reserves were brought up, having no Old Guard to meet them, the tactical advantage was theirs and not ours—we were repulsed.

Tactical success in war is generally gained by pitting an organized force against a disorganized one. This, at least, is one of the secrets of Napoleon's success. At Ypres we had not sufficient means to disorganize the enemy; at Cambrai the enemy did not offer us the opportunity of disorganizing him; both battles were, in my opinion, conceived on fundamentally unsound tactical premises. What we now want to aim at is a combination of the above two ideas:

(i.) To force the enemy to mass his reserves in a given area.
(ii.) To disintegrate these reserves before we attempt to annihilate the enemy.

This done, pursuit—that is, the tactical act of annihilation—becomes possible. Pursuit produces the dividend of battle.

The more reserves we can force the enemy to mass, as long as we can disorganize them, the greater will be the tactical interest on our outlay. This is the crucial problem of the offensive. This is why Napoleon said: "There are many good generals in Europe, but they see too many things at once. I seek the enemy's masses in order to annihilate them." In applying the principle of offensive action we must not be misled into seeking merely for a weak point, but for a vulnerable point at which we may attack the enemy's vitals. The difference between guerilla warfare and *la grande guerre* is that, whilst in the former we strike at packets of men, or individuals, in the latter we strike at organized forces under a central command. Do not let us delude ourselves into supposing that because the enemy's reserves are not at hand it is the time to attack. It may be the time to attack, if with those reserves the probabilities are that he will defeat us; but, if otherwise, it may be the very worst time to do so. "*Qui ne risque rien n'attrape rien*" was a favourite saying of Napoleon's. The mainspring of the principle of offensive action is audacity—that is, exalted determination to win.

## 21. The Object of the Offensive

In chapter viii. I defined the grand tactical object of battle as being "the destruction of the enemy's military strength as represented by his command and organization." Though this object remains stable, the tactical objectives vary with conditions and the means of action at the disposal of the general.

In the past these tactical objectives have been gained by destroying the enemy's field armies; but, as I have explained, the potential strength of a body of men depends on the maintenance of its organization. If this organization is destroyed we have destroyed its strength, and so have accomplished our object.

There are two ways of destroying an organization:

(i.) By wearing it down (dissipating it)
(ii.) By rendering it inoperative (unhinging it).

In war, the first comprises the killing, wounding, capturing, and disarming of the enemy's soldiers (body warfare); the second, the rendering inoperative of his power of command (brain warfare). Taking a single man as an example, the first method may be compared to a succession of slight wounds which will eventually cause him to bleed to death, and the second to a shot through the head.

The brains of an army are its staff—army, corps, and divisional headquarters; could we suddenly remove these from an extensive sector of the enemy's front the total collapse of the fighting personnel would be but a matter of hours, even if only slight pressure is exerted against it. Suppose, now, that no pressure is exerted, but that, in addition to the shot through the brain, a second shot is fired through the enemy's stomach—that is, his supply system behind his protective front; then his men will either starve to death or disperse to live. The fact I wish to accentuate here is that, as our present theory of offensive action is based on the idea of destroying personnel, new means of war, so I am convinced, will force us to substitute a theory based on the idea of destroying command—not after the enemy's personnel has been disorganized, but, when it is possible, before it has been attacked,[1] so that it may be found in a state of disorganization when attacked. I am convinced that this will take place, because in this form of attack I see the highest application of the principle of surprise—surprise by novelty of action—or the impossibility of counter-action even when the unexpected has become the commonplace.

Novelty of action in its turn demands novelty of means. The means are movement, weapons, and protection; consequently, if in the attack military force is to be economized, these means must be superior to those of the enemy. Though it is through mind that the principle of offensive action is applied, its means of expression are movement, weapons, and protection, and, if these means be obsolete, though the principle does not change its application may become impossible. If, on the other hand, these means be vastly superior to the enemy's, then an intelligent application of this principle may produce immediate and overwhelming success.

[1] We are apt to despise the Bolshevist armies and military operations, but we have much to learn from them, for their leaders are as unshackled by rules, regulations, and traditional methods of war as were the Revolutionary generals of 1792–97. Before a physical attack was launched on Kolchak and Denekin the areas occupied by these generals were morally attacked by propaganda. Their base of operations was thus undermined, and their power of command shaken severely, before the general attack was launched.

# CHAPTER XIV

## THE PRINCIPLES OF RESISTANCE

By restless undulation; even the oak
Thrives by the rude concussion of the storm.
—COWPER.

### 1. THE PRINCIPLE OF DISTRIBUTION

THE principles of resistance form the base of the principles of pressure, and the relationship between them is expressed by the principles of control which regulate the expenditure of force; consequently, if force is to be expended economically expenditure will depend on the correctness of our resistance, which is governed by the principles of distribution, endurance, and security.

In war-time endurance is immediately affected by danger, and, fear being aroused, the natural inclination of the soldier is to secure himself against it. This desire to seek protection reacts on the determination of the commander, and frequently compels him to distribute his troops in such a manner that pressure cannot be exerted to the full.

In peace-time danger is absent, consequently soldiership is endowed with a pseudo-courage which leads to an unreal application of the principle of distribution of force. I shall revert to this subject when I examine the principles of endurance and security, but I mention it here because, when danger is absent, nothing appears easier than to distribute our forces correctly, whilst in fact, on account of this absence, it is a most difficult problem. In brief, the problem of distribution is as follows:

We first decide on our object, whether it be the winning of a war or the capture of a sentry-post. To gain this objective demands an expenditure of mental force directed against an enemy probably as strong willed as ourselves. We know from the study of human nature that, if we can unhinge the enemy's *moral*, we shall weaken his fighting power, consequently we seek how to surprise him. Our projected direction now becomes coloured by this intention, namely to take him at a disadvantage. If we can surprise our enemy we shall economize our fighting power, and particularly our *moral*. Surprise is, therefore, of immense economic value ; consequently, if force is to be distributed correctly, our distribution must not only aim at effecting surprise, but of countering it by endurance. The distribution of force is firstly a problem of *moral.*

Next we secure ourselves against attack, and, by applying the principle of security, we establish a solid base to work from. Our maximum security will be attained when the enemy is defeated ; our maximum effort must, consequently, be directed towards concentrated offensive action, and the less material we use up in building our foundations the more we shall have in hand for the superstructure. Here, then, are two problems. Out of a given force, what proportion of this force should we use for the foundation of the operation we contemplate, and what portion for the operation itself ? The answer to these two questions is arrived at by applying the principle of distribution in accordance with the conditions of war.

### 2. The Dependence of Distribution on Conditions

Of all the principles of war, the principle of distribution of force is the most difficult to apply, because of its close dependence on the ever-changing conditions of war. Economy does not mean storing up, but expending wisely, and expenditure demands distribution, since conditions are always changing. Our total force is calculable in any set of circumstances, if the nature of these circumstances is known. But they seldom are known, and they are perpetually changing ; nevertheless, the side which can evalue the conditions of war the more correctly is the side which can apply this principle more fully. Certain conditions surround us as to the value of which there should be little doubt, and one of the most important of these is the *moral* of our men. To economize the moral energy of his men a commander must not only be in spirit one of them, but he must ever have his fingers on the pulse of the fighters. What they feel he must feel, and what they think he must think. But, whilst they sense fear, experience discomfort, and think in terms of victory or disaster, though he must understand what all these mean to the men themselves he must in no way be obsessed by them. To him distribution of force first means planning a battle which his men can fight, and, secondly, adjusting this plan (the mental factors) according to the physical and moral changes which the enemy's resistance is producing in their endurance without forgoing his object. This does not only entail his possessing judgment, but also foresight and imagination. His plan must never crystallize, for the energy of the firing-line is always fluid. He must realize that a fog, a shower of rain, a cold night, an unexpected resistance, may force him to readjust his plan to the change in conditions, and, in order to enable him to do so, adjustments in distribution depend on his reserves, which form the staying-power of the battle and the fuel of all movement.

### 3. Economic Distribution of Force

Before the strength of a reserve force can be decided on it is necessary to work out a provisional distribution of force. We have decided on our object, and we have agreed, I will suppose, to surprise our enemy by moving against his left flank. We have also considered the most probable moves that the enemy is likely to make, and have temporarily decided that a certain portion of our force must be earmarked to secure our attack, and, if this attack succeeds, that we must follow it up by a pursuit, and, if it fails, that we must either reinforce it, attack in another place, or cover the withdrawal of the attackers. To begin with, we must distribute our total forces in three categories:

(i.) Protective troops.
(ii.) Offensive troops.
(iii.) Reserves (including troops for the pursuit).

The next question is, How are we to decide on the strengths of these forces?

We must turn to the conditions of war—the enemy, the theatre of war, communications, and time. There are many other conditions, but these four will suffice for my present purpose. We know approximately the enemy's strength, approximately his position, but very seldom his intentions. We can, however, step into his shoes, and, giving him full credit for common sense, we can work out a plan for him. From a good map we can study the theatre of war and the communications contained in it. We can divide it into areas which will resist movement and areas which will facilitate it, and then with a pair of dividers and a time scale we can consider our distribution.

The duties of our protective troops may be one or more of the following:

(i.) To screen the advance of the offensive troops.
(ii.) To protect them before, during, and after battle.
(iii.) To protect communications and bases.
(iv.) To restrict the enemy's movement in certain areas,

The duties of the offensive troops are:

(i.) To attack or counter-attack the enemy.
(ii.) To threaten the enemy or his communications and force him to form detachments.

And those of the reserves:

(i.) To maintain offensive or protective strength.
(ii.) To maintain freedom of manœuvre.
(iii.) To effect concentration of force.
(iv.) To meet unexpected situations.
(v.) To carry out the pursuit.
(vi.) To cover a withdrawal after a reverse.

From the above it will be seen that numerically the duties of the protective and reserve forces are greater than those of the offensive ones. This does not necessarily give us any fixed measurements of protective, offensive, or reserve strengths, but it does hint that until we actually engage the enemy our protective strength should be strong, our offensive strength weak, and our reserve strength as strong as possible, because it is from our reserves that we feed our offensive and protective operations. In an encounter battle, or one delivered against a defensive position, first we want to limit the enemy's freedom of movement, either by resisting him or pinning him down—the physical attack; secondly, we want to surprise him—the moral attack; and thirdly, to drive this surprise home and overwhelm him—the decisive attack. When we study military history we shall find that two initial faults are always recurring. The first is insufficiency of initial protective power, and the second insufficiency of reserves. The object in war is not normally gained by an initial offensive in strength, but by an initial resistance under cover of which genius can gain its end by a skilful use of reserves—in other words, by an economical distribution and utilization of force. The bull generally succumbs to the skill of the matador; this is not a principle of war, but a very good rule to remember.

On the battlefield itself, to economize our own strength and to force the enemy to dissipate his by means of feint operations and surprisals is the first offensive step towards victory. Every weapon which we can compel the enemy to withdraw from the point of attack is an obstacle removed from the eventual path of progress. Every subsidiary operation should be related to the object, and effect a concentration of force on the day of decisive action. Every subsidiary operation should add, therefore, an increasing value to final victory—that is, the power of producing a remunerative tactical and ultimately political dividend from the force expended during the war. Thus, even in so small an operation as a raid executed by twenty men the question must first of all be asked, What will be the tactical dividend if the operation proves successful? Will five per cent. be a sufficient recompense or should the action produce ten per cent.? "Is the game worth the candle?" This is the question every commander must ask himself before playing at war.

By this I do not mean that risks must never be taken; far from it. It is by taking risks *which are worth taking* that, more often than not, the greatest economies are effected and the highest interest secured. In war, audacity is nearly always right, but gambling is nearly always wrong, and the worst form of gambling is gambling in small stakes; for by this process armies are bled white.

### 4. The Relationship between Defensive and Offensive Distributions

By now I trust it will be realized that economy of force is gained by distributing force economically. I have stated more than once that, though in theory we may find it easier to think of actions as possessing offensive characteristics, in practice we must think in one term—*the protected attack*—whether we are advancing, retiring, or standing still. Such an attack is the relationship between protective and offensive power, and this relationship is governed by distribution.

From the standpoint of the defensive, protection is gained by shielding; it is but a means to an end, the end being victory and the means being life. Living men win battles, and, the more highly armed living men we can bring on to the battlefield and maintain there, the greater will be our chances of victory. Therefore whatever reduces living men to dead men must be secured against.

From the standpoint of the offensive, protection is gained by striking out, and striking out not only requires living men, but men who can give blows. The more blows we give the less we shall receive; for our opponent, being reduced to shield himself, will possess less means and opportunity to strike at us. Given a sword and a shield, a man will, when threatened, simultaneously raise his shield and draw his sword. The shielded attack is uppermost in his mind. To him it is instinctive protection to kill his adversary. With masses of men it is the same; the surest protection is the elimination of the danger,

From the above we can extract the following facts:

(i.) The offensive is the strongest form of the defence.
(ii.) The defensive is but a suspended state of the offence.
(iii.) The offensive requires every available weapon so as to transmute the enemy's offensive into a defensive.
(iv.) The defensive requires only sufficient men to maintain and protect the offensive.
(v.) The offensive, being dynamic, requires the highest ability, dexterity, and power of movement.
(vi.) The defensive, being static, requires skill with less mobility, and determination without a high degree of innovation.

From these facts may be elaborated the following theory:

(i.) The offensive should be assumed on all occasions in which circumstances permit of it.
(ii.) The defensive should be so organized as to permit of it changing into an offensive at the shortest possible notice.
(iii.) The offensive cannot be too strong (endurable), therefore the defensive should not employ a weapon beyond the number absolutely necessary for security.
(iv.) The offensive will require masses of weapons, consequently every weapon that can be spared from defensive work should be held in hand for offensive action.

This theory, I think, is based on sound reasoning, therefore to discard it is an act both dangerous and foolish, unless the ruling conditions are such as to render the principle of offensive action inoperative. In this case the most obvious thing to do is to cry quits or abandon the war and crave peace; so that before complete destruction supervenes—and this is what passive defence leads to—war may be terminated and the offensive resumed at some later date, when circumstances are more auspicious.

Nevertheless, if the pages of history be consulted it will be discovered that this theory has been subjected to many a rude shock, and to the detriment of the infringer.

The following, drawn from the past, are errors worth remembering:

(i.) The offensive languishes on that side which is least prepared to wage war, and which is, through ignorance of the principles of war, blinded by the belief that the enemy must be held back at all points; and that consequently it is necessary to be everywhere equally strong in men and superlatively strong in defences.

(ii.) The neglect of peace teaching, based on the experience of former wars, generally leads to the creation of "impregnable positions," in place of such preparations as will aid a rapid assumption of the offensive.

(iii.) The all but total depletion of a reserve—that is, a striking force—on account of the stringing out of troops for purely defensive tactics, such as the passive holding of trenches, villages, and fortified positions, renders a sustained offensive impossible.

(iv.) The general demoralization and disorganization of all ranks by the incessant creation of new defences, and the repair of old ones, detrimentally affects training and leadership, and consequently lowers the offensive spirit of all concerned.

From what I have said I hope it will be realized that, in practice, there is no dividing-line between the offensive and defensive in warfare, and, if an artificial one is created, correct distribution will not result. In offensive or defensive operations the object is identical. The object of the defensive (shielding) is not merely to preserve our lives, but to preserve them so that *we may more economically destroy the enemy's strength.* Consequently a defensive battle is based on an offensive plan or idea, which, through force of circumstances, cannot at once be put into operation.

Superiority of weapons at the decisive point means superiority of offensive power, and lack of this superiority is frequently the direct cause of defensive action. If men are squandered in attempting to avoid blows they will not be in a position to give them, and, not giving them, they allow their enemy to reduce his defensive strength to a minimum and to increase his striking power in proportion. It was against this type of warfare that the great Napoleon inveighed when he wrote to his brother, the King of Spain, saying: "The cordon system is only good against smugglers."

In order to obviate the inherent disadvantages and vices of the cordon system, the theatre of war, area of operations, or battlefield must be divided into positions of resistance and lines of pressure. These must be chosen from the point of view of the grand offensive, and all the stages of the offensive must be based on these positions.

For those detailed to resist the enemy, their immediate object is not to defend the position occupied, but to aid the offensive, whether this offensive be next door to them or hundreds of miles away.

The cordon system simultaneously infringes the principles of distribution and of concentration, for, the defensive being the aim of this system, a time arrives when the offensive becomes inoperative, not through lack of weapons, but through impossibility to concentrate them, due to their faulty distribution.

The strength of garrisons must be in proportion to the defended areas they are ordered to occupy. Ten men will hold a blockhouse, and a blockhouse may delay a brigade; ten men will not hold a fortress; therefore, in our defensive plans, do not let us build fortresses when blockhouses will suffice. The strength of defences does not lie in their size, but in the harmony between their size and the strength of their garrisons.

### 5. The Relationship between Distribution and Movement

Distribution of force is also closely related to economy of movement. Many generals have attempted to win a Marathon race in sprinting time; they have thrown in all their reserves at once, and have lost their wind a few hours after the battle has begun. Such operations as these are doomed to failure long before the first shot is fired. Others, through an over-extension of troops, particularly in those employed in protective and defensive duties, have found it most difficult to build up a reserve when such a force is required, time being insufficient to carry out the necessary concentration. Consequently, before we plan our defences we should consider the following maxims:

(i.) When from a state of defence the offensive is assumed, this act should in no way disorganize the existing defensive arrangements.

(ii.) Any delay in the assumption of the offensive from the defensive may prove fatal to both operations.

(iii.) In offensive action, *moral* weakens in proportion as improvization increases.

The lesson which these maxims teach is the vital necessity of a strong reserve in order to supply an army with motive power. If an economical distribution between offensive and protective troops has been made, it should normally be unnecessary to switch the protective troops on to offensive work. In place, the extra offensive power should be drawn from the reserves, and the protective troops no longer required should be relegated to the reserve. In a prolonged campaign, if the principle of distribution is to be maintained, it is just as necessary to feed the reserves as to feed the firing-line. I noted this when I examined the principle of mobility; here I will only point out that economy of distribution is frequently affected by this principle, because, if the distances between the various parts of an army are great, or the means of movement slow, though offensive action may at first succeed, it will be found impossible to maintain a sufficient reserve to keep offensive action fluid.

### 6. Distribution and Weapon-Power

Distribution of force is also directly affected by the losses incurred. Every man killed or seriously wounded must be replaced, not only to maintain sufficiency of strength, but to maintain tactical organization. Whilst power of action must be kept fluid, organization, as far as possible, must be kept stable, for a fluid organization is a bad base for activity to work from. The continual replacement of casualties by men drawn from the reserves detrimentally influences organization and hampers leadership and command. In the recent war all the contending parties were so imbued with the idea that resistance was the main operation of war that thousands of men were slaughtered in order to hold a few miles of tactically valueless ground. Frequently during the years of position warfare the tactical value of positions was entirely lost sight of, and replaced by the idea that no position occupied must be evacuated; this idea being particularly comforting to incompetent commanders who were incapable of redistributing their troops. Needless to say, the value of a position depends on the tactical and strategical conditions which surround it, and if in attempting to secure one man behind a shield we lose ten shield-bearers, all good fighters, the operation is obviously an uneconomical one; yet this faulty distribution was constantly being made during the Great War, because the true purpose of the shield was not understood.

In offensive actions the losses were appalling, and this undoubtedly forced defensive operations to the fore. Only during the last year of the war on the Western Front was a more economical distribution established between protection and offensive action, and this was almost entirely due to a comparatively small number of armoured machines which enabled mobility and concentration of force to be applied. Though statistics are frequently misleading, the following are at least of some interest:

(i.) From July to November 1916 the British Army lost approximately 475,000, it captured 30,000 prisoners, and occupied some 90 square miles of enemy country. The casualties totalled to 5,277 per square mile.

(ii.) From July to November 1917 the losses were 370,000, the prisoners captured 25,000, and the ground occupied was about 45 square miles. The casualties per square mile were 8,222.

(iii.) From July to November 1918 the losses were 345,000, the prisoners captured 176,000 and the ground occupied was about 4,000 square miles. The casualties per square mile were 86.

Whatever the reasons for this reduction of casualties may have been, they should be discovered, so that we may learn more about the conditions which compel us to expend force, and the conditions which enable us to economize this expenditure.

Before and during the war all sides were obsessed by human tonnage. A study of the second book of Xenophon's *Cyropædia*, I think, might have disillusioned them, and brought them to realize the influence of superior weapon-power on the principle of distribution.

"I see," said Cyrus, "you reckon our cavalry at less than a third of the enemy's, and our infantry at less than a half."

"Ah," said Cyaxares, "and perhaps you feel that the force you are bringing from Persia is very small?"

"We will consider that later on," answered Cyrus, "and see then if we require more men or not. Tell me first the methods of fighting that the different troops adopt."

"They are much the same for all," answered Cyaxares, "that is to say, their men and ours alike are armed with bows and javelins."

"Well," replied Cyrus, "if such arms are used, skirmishing at long range must be the order of the day."

"True," said the other.

"And in that case," went on Cyrus, "*the victory is in the hands of the larger force; for even if the same number fall on either side, the few would be exhausted long before the many.*"

"If that be so," cried Cyaxares, "there is nothing left for us but to send to Persia and make them see that if disaster falls on Media it will fall on Persia next, and beg them for a larger force."

"Ah, but," said Cyrus, "you must remember that, even if every single Persian were to come at once, we could not outnumber our enemies."

"But," said the other, "can you see anything else to be done?"

"For my part," answered Cyrus, "if I could have my way, *I would arm every Persian who is coming here in precisely the same fashion as our Peers at home, that is to say, with corslet for the breast, a shield for the left arm, and a sword or a battle-axe for the right hand.* If you will give us these, you will make it quite safe for us to close with the enemy, and our foes will find that flight is far pleasanter than defence. But we Persians," he added, "will deal with those who do stand firm, leaving the fugitives to you and your cavalry, who must give them no time to rally and no time to escape."

That was the counsel of Cyrus, and Cyaxares approved it. *He thought no more of sending for a larger force, but set about preparing the equipment he had been asked for. . . .*

Two thousand four hundred years ago it was recognized by the clear-sighted Xenophon that victory is not to be sought in distributing or concentrating *masses of men*, but in *perfection of weapons*. Weapon-power and *moral* are the two greatest sources of battle energy. Xenophon realized this, and he understood the principle of distribution of force, and the conditions in which this principle *could* operate, better, far better, than did any general in any European army in 1914. What a lesson! Two thousand four hundred years old, and we have not learnt it yet!

### 7. The Principle of Endurance

Distribution as a mental principle governs the moral and physical spheres as far as resistance of moral and physical force are concerned, as I have just shown by a quotation from Xenophon in which we see Cyrus thinking out the more economical distribution in terms of weapon-power and *moral*. He realizes their intimate relationship. If all soldiers are equipped in a similar manner, and two armies engage in battle, then, if other things be equal, the numerically stronger side will win, because it is able to concentrate superior force on the battlefield. He might have attempted to rectify the inequality of the Medes by proposing higher generalship, but this is not a commodity which can be bought, and it may take a generation or more to cultivate it. He might have proposed imbuing the Median soldiery with a fanatical courage, but again this demands a slow process of education or the rapid process unconsciously applied by some religious genius. No, time is short, and it generally is so in war, so in place he argues: Our men are human, and they are possessed by a will to live and a will to fight; if I can only increase their means of fighting, so that they are superior to their enemy's, then in inverse proportion will danger be reduced, and, as it is reduced, so in direct proportion will their *moral* be increased, and, as *moral* rises, so will their determination to conquer grow.

This I believe to be the inner meaning of the dialogue between Cyrus and Cyaxares, and it is for this reason that I concluded my brief survey of the principle of distribution with this quotation; for it closely links the principle of distribution to that of endurance, and, by showing how *moral* can be safeguarded and cultivated through physical means, it also links the principle of endurance to that of security.

The principle of endurance, in my opinion, is the principle which, under one name or another, has been most discussed in modern times, but least understood, and so misunderstood that the unnecessary waste of life resulting has been truly appalling. *Moral* has been on everyone's lips, and even the last joined subaltern will freely talk of the *moral* of his men as if it were a commodity. If a man is singing he says his *moral* is high; if grumbling, that it is low; discipline, obedience, cheerfulness, are all mistaken for *moral*, which, in fact, as I have shown, is a form of self-sacrifice. It is the artificial cultivation of an instinct in order to balance a higher or more potent instinct—that of self-preservation. This balancing process depends on the influence of the moral conditions of war on the instinct of self-sacrifice, which, like every other instinct, must be brought under the dominion of the will, if the will is to be a free agent. The principle which governs these influences is the principle of endurance.

### 8. The Influence of Physical Conditions

Our outlook upon endurance has been alchemical. We all have realized the influence of physical fatigue on the moral condition of our men. We all know that an exhausted soldier is a bad fighter. But, whilst we generally attempt to bring our troops on to the battlefield physically fresh, once there, we expect their endurance to continue erect like a material target until it is knocked out. We realize what physical strain means to *moral* before battle, but we do not realize what the strain of the battle itself means to *moral* until this strain begins to exert its sway, when it bends up the endurance of an army like a tornado striking a forest.

The reader may say this is ridiculous, and that we do realize it. I answer we do not, and in proof I urge that the reason why the Great War of 1914–18 was mainly a static operation on *all* fronts was because the offensive pressure of modern firearms was too much for moral resistance in the open, and that the terror of death could only be rendered endurable by going to earth. The soldier had, in fact, to encase himself in earth, like a limpet in its shell, in order to hang on to the rim of the battlefield. In brief, the unarmoured (either by earth or steel) man *will not*, in ninety-nine cases out of a hundred, face the machine-gun in the open.

In 1914 the maximum aimed fire of a division of infantry in line was about 50,000 rounds a minute; to-day, on account of the enormous increase in automatic weapons, it is about 150,000 rounds. Will it be contended that the moral endurance of our men is three times as high as it was in 1914? If it is, then the operations of 1914 are likely to repeat themselves. But this cannot possibly be urged, since human nature remains approximately constant; therefore the conclusion is that, if war broke out to-day, it would in character be even more static than it was ten years ago.

As the war proceeded, from the occupation of fixed earthworks, the soldier got into mobile steel-works—tanks and armoured cars—and, as armour enabled Cyrus to win his battle, so once again did armour enable such battles as Cambrai, Soissons, and Amiens to be won. The human nature of the soldier was the same, whether in a trench or a tank, but in the tank physical security safeguarded *moral* and, consequently, it could endure, and supply moral armour to the soldier's will, his determination to win, by instilling fear into his adversary, which fear was potentized by the fact that the enemy's infantry were impotent against tanks.

To-day, we have tanks, at least a few, but we still rely mainly on infantry, and, as I stated in chapter I., we still believe that infantry is the superior arm, and, as our belief is not founded on fact, it is for this reason that I maintain that our outlook on endurance is alchemical. As long as we have faith in this belief, then, whatever we may think we can do during peace training, we shall suddenly know that we cannot do in our next great battle, and, through our inability to apply the principle of endurance, any attempt to apply the remaining eight principles of war must suffer, for true economy of force is unobtainable unless all are applied.

### 9. The Influence of Weapons on Discipline

Discipline is the mental, moral, and physical system applied to prepare the soldier for war, and, as he cannot possibly fight in any other war but the next one, the controlling factors are the conditions of this war.

The next war will be evolved from the existing conditions of peace; thus the nature of peace will give us a clue to the nature of the next war. I do not intend to enquire into this condition as it faces us to-day, but it should be realized that as every advance in civil progress demands a commensurate advance in the progress of war, so does every advance in the physical means of waging war demand a change in military discipline.

In the days of Frederick William of Prussia the kingdom he ruled was still in a feudal, certainly more than semi-feudal, condition. The national outlook was aristocratic, if not exactly autocratic; the masses of the people were looked upon by the dominant class as higher animals, the dominant class alone was human. The rank and file of the army were recruited from the lowest stratum of society, and commanded by the higher strata. The difference between leaders and led was, consequently, one of superiority and inferiority; the rank and file being looked upon as mere cannon-fodder. The moral outlook on war was pre-eminently brutal.

The tactics of the day added to this brutality; the lack of intelligence in the rank and file and the precision of the mechanical tactics of the day demanded an unthinking human machine which could approach an enemy to within fifty paces, and then load, present, and fire in so accurate a timing that years of drill were required in order to attain perfection. As the men were looked upon as animals, and as they possessed little intelligence and practically no sense of patriotism, discipline was instilled through fear. No appeal was made to heart or brain, but if a fault occurred a man's back was lashed bare to remedy it.

Frederick the Great won all his battles with the cat-o'-nine-tails, and his system was possible, since in a closely set three-rank line, no initiative save that of the commander could be developed. This human wall moved to the voice of one man, which, unthinking, was obeyed, since it commanded more terror than the enemy.

The eighteenth century was a period of decadence, and decay is the herald of growth. Out of the materialism of the period was struggling forth a new spirit—the spirit of humanity, which at length found expression and revenge in the French Revolution. Society was upheaved, and so was the art of war. Years of drill were now impossible; command by brutality was frustrated by insubordination. The French Revolutionary armies were untrained, and, lacking discipline, instinct took control, and the soldier, lacking the authority of command, fell back on his intelligence, and through native initiative and cunning sought to protect himself by skirmishing or by quitting the battlefield. Thus it was that the rigid wall was replaced by mobile fragments, which by degrees took form, grouped themselves, and were known as the *voltigeurs* of France.

In spirit the old system had gone, but as a shibboleth it

lingered on in all countries outside France, and, lingering on, brought to the front a small number of rational and courageous men who saw that discipline demanded a new spirit, the spirit of loyalty and affection.

Sir John Moore saw this quite clearly in England; he saw that the weapon of the day demanded tactics which permitted of an extended order of fighting; he saw that an extended line of men, or of groups of men, could not be commanded through fear, but only through affection. If between officers and men a family spirit could be established, then, as a son will fight for his father, so will a private soldier fight for his officer. He introduced, therefore, a new discipline to fit the new tactics, a discipline which was not based on fear, but on affection, not on the instinct of self-preservation, but of self-sacrifice, and to-day his system is the system of the British army and of most foreign armies as well.

In the days of Moore, Napoleon, and Wellington, the line, as the tactical attack formation, still held its sway, but by the end of the century it had become so elastic that in the South African War of 1899–1902 we find extensions of as much as fifty paces between men in the firing-line. To command such a line, even when the men were devoted to their officers, was most difficult, since the line could not be led—its length prohibited this—and since each man was isolated, and, not having been trained to fight on his own, lost confidence in himself. Thus it happened that the more intelligent, or rather less ritual-shackled, Boer frequently defeated us.

As the improved musket and early rifle in Moore's day had forced a change in discipline, so the magazine rifle demanded an equivalent change. It demanded of the soldier intelligence in its use as well as affection for his leader. But this change was not observed, consequently the endurance of the soldier was not understood.

By the date of the declaration of war in 1914, twenty years after the introduction of the magazine rifle, the discipline of the Martini-Henry, the Chassepôt, and the Needle gun still held sway. Leaders had been highly trained, but those who followed them had not been taught to lead themselves. The machine-guns decimate the most extended lines, and even the most loyal and self-sacrificing troops in the world were reduced to impotence, because, once their leaders were killed and wounded, they could not lead themselves.

The manipulation of the machine-gun, especially in mobile warfare, demanded team-work; the use of tanks to blaze trails through the enemy's entanglements demanded the close support of small packets of infantry, and so did shell-hole fighting. The tactical formation of the line of individual fighters was thus changed into a line of small packets of fighters, each packet operating at a considerable distance from the next. As distance has increased, so must intelligence increase, for every packet is a minute army which must hit, guard, and move in a definite area some two hundred yards in width. The leader of each packet has got to fight his own battle, as well as co-operate in the general battle, he must consequently be a man of high intelligence and determination. If he is killed, one of his followers must replace him, if continuity of movement is to be assured. Therefore all his followers must be intelligent and determined men, so that, if all become casualties save one man, this one man may continue to press on and co-operate with the groups on his flanks. Each man must be so disciplined that his endurance is based on fear, on affection, and on intelligence. He must be afraid to run away, because he will be punished—the endurance of Frederick; he must be willing to push on, because he has a high *esprit de corps*—the endurance of Moore: and he must have the intelligence to apply, in his own small sphere of action, the principles of war, because unless he can apply them he cannot fight intelligently.

I have gone to this length to show the influence of weapons on discipline because I am convinced that to-day it is one of the most important of military problems, and that, unless we reform our discipline, we shall never, in existing conditions, and still less so in those which are likely to confront us in the next war, possess that *moral* which will permit of us applying the principle of endurance of force.

## 10. Mental Discipline

The influence of weapons on discipline is only one of many of the conditions of war which determine the moral endurance of the soldier. Two other important series are those which include means of movement and protection; others are education, civic-sense, social outlook, etc., etc.; but most of these lie outside the sphere of immediate military control. The example I have taken must, therefore, be considered simply as an example, and not as the only example, and, in place of multiplying examples, I intend briefly to examine the threefold order of discipline, namely discipline in the mental, moral, and physical spheres.

In the mental sphere I have already more than once accentuated the importance of knowledge, understanding, and wisdom. The importance of these qualities of mind is catholic and not sectarian, for, in his own sphere of action, it is as important for a private soldier to be knowing, understanding, and wise as it is for the general-in-chief in his. In both ignorance is a bane and a curse, as it is in all spheres of life. Of ignorance Mr. Gore writes: "... it is those who know not what to expect who experience the most anxiety. Ignorance, fear, and terror go together. ... Ignorant persons fear intelligent ones, because they dread lest the powers which knowledge confers be used to their injury." And again: "There are various other symptoms of ignorance, and amongst them are—indecision and fear of the natural risks of life. By paralysing the will through deficiency of sound ideas, ignorance causes indecision and want of promptitude, or else it makes men reckless from sheer desperation; without suitable knowledge a man cannot act safely or promptly."[1]

These quotations refer more particularly to ignorance in everyday life, but they are as applicable to the soldier as to the civilian. The soldier who does not know what war entails, when surrounded by war conditions, fails to understand them; he is surrounded by a fog of ignorance, fear is magnified through this mist, and reality, which he cannot understand, becomes a mirage of false dangers. Not being able to see consequences, he not only cannot see ahead, but cannot look around; he is blind and full of fears.

Knowledge, understanding, and wisdom, the three qualities which beget mental endurance, are not to be sought on the field of battle, and, unless mental discipline has been cultivated and ceaselessly cultivated during peace-time, it can seldom be cultivated during war, and then, as I have already stated, only at tremendous cost. To cram facts into our men's heads (the normal process of education) is not sufficient, for we must fashion our mental discipline so that they themselves can cultivate understanding. To understand requires examination; it requires criticism. On the battlefield we all have to obey someone, and generally spontaneously, but in peace-time it is different, and the intelligent man, whether soldier or officer, should be allowed to say: "I do not like the plan you suggest. I consider that it should be done this way." Then let both ways be tested and compared, for in their differences is to be sought true knowledge and understanding.

In war we are faced by an enemy; in peace the enemy is ourselves; it is through encouraging others to criticize our ideas and actions that we attack ourselves and discover our errors. To-day mental discipline is all but unknown, and consequently

[1] *The Scientific Basis of Morality*, G. Gore, pp. 392, 413.

obedience is blind, and men enter war blindfolded. On the battlefield action demands spontaneous obedience, therefore, if during peace-time we have cultivated true mental discipline, in war we shall move forward with our eyes open.

### II. Moral Discipline

Moral discipline is not only based on those sentiments which stimulate the instinct of self-sacrifice, but on a knowledge of the conditions of fear. Knowledge in the moral sphere is as important as knowledge in the mental. Affection is a sacred quality, and not one which should be prostituted. Hitherto it has largely been attained by providing physical comfort—by interior economy, good feeding, clean and pleasant surroundings, etc., all of which are admirable, but not sufficient; for, after all, a normally moral man will provide such for his dog. What we must aim at is to superimpose on all these excellent conditions a moral discipline based on respect. I have touched on this subject in chapter xii., when I examined the principle of determination, but there I dealt mainly with the general-in-chief; here my concern is with all officers and leaders. This respect is based largely on the intellectual and moral qualities of the officer; is he worthy of a man's affection and awe.

> An ignorant person is rarely highly moral; first, because it requires knowledge to enable us to do unto others as we would have them do unto us; second, because, in the numerous difficult cases which occur with all men in going through life, an ignorant man is often unable to determine what is right; and third, because it requires knowledge and reasoning-power to predict the consequences of our acts, and to distinguish truth from error.[1]

An ignorant man cannot be a good soldier. He may be brave and audacious, and, in the hand-to-hand struggles of the past, his ignorance may have appeared but a small defect, since he could rapidly clinch with danger. But to-day this defect has grown big; the stout arm of Cannae, of Crecy, or even of Inkerman, demands at least a cunning brain. Fighting intervals and distances have increased, and there is more room for ignorance to display its feathers, and the corridors of fear are long and broad.

To-day, unless the soldier understands the realities of war, unless he understands what is going to make him fearful and how he is going to turn this condition to his advantage by making his enemy more fearful than he is, he opens himself to vigorous

[1] *The Scientific Basis of Morality*, G. Gore, p. 399.

surprise, and, even if he overcomes this surprise, his economy of force must suffer.

One of the most damnable of heresies is to suppose that, if we keep the soldier in ignorance of the realities of war, such as the power of the machine-gun, the power of the tank, of gas, etc., we are going to shield his *moral* on the battlefield, because he will step on to it unconscious of danger. This heresy belongs to the Satanic creed of ignorance. Ignorance is not only *always* wrong, but it is *the* evil of the world. It is not by ignorance that we stimulate the endurance of our men, for it is by knowledge and understanding of the realities of war that we do so. This understanding, by fortifying courage, strengthens determination, which, coupled with wisdom, leads to economy of means and of action.

### 12. Physical Discipline

Physical discipline is discipline of the body over the means—the weapons, means of movement and protection, and the employment of these means in harmony with the most likely conditions in which they will be used. This discipline aims at economizing, through a correct use of means, the expenditure of moral force so that the will of the general can express itself more fully.

The main fault in existing physical discipline lies in a lack of appreciation of the true meaning of moral endurance, and its cultivation. We are still guided by the shibboleths of the Prussian System. Thus, though we should now have realized the importance of stimulating individual initiative, we cramp this quality by months of close-order drill, which, in place of developing it, induces a comatose collective spirit which has no will of its own.

In the days of Frederick the Great, as I have shown, the unthinking instrument was at least an effective weapon, since the voice which commanded obedience on the drill square could equally well command it on the battlefield. To-day this is no longer possible, consequently our aim should be, not to drill our men into unthinking machines, but, instead, to cultivate within each one of them a high sense of leadership. If we were to spend as much time in training leaders as we now do in creating automata, we should certainly gain in the physical discipline which the modern battlefield demands, even if our men lost some of that antiquated elegance which is so attractive on the parade ground.

Leadership cannot be taught as a drill, for leadership, like dry-fly fishing or riding a horse, does not depend anything like so much on book knowledge as on discovering one's own limitations and on overcoming self-consciousness. To train our men to become leaders, we must allow them responsibility, for it is through responsibility that leadership is cultivated. A child responsible for the care of a hutch of rabbits will cultivate a higher sense of leadership than a full-grown man bellowed at by another on the drill square.

As regards weapon-training, which should be included in the category of physical discipline, I will not say much, as its value is universally recognized. Yet one point is frequently overlooked, namely that, though each weapon possesses certain definite powers, these in battle are modified by the power of the other weapons; consequently, unless we understand this correlation and train accordingly, the conditions of our weapon-training will not coincide with those experienced in war.

I think that I have now shown the complexity of the various conditions in which we are called upon to apply the principle of endurance, a principle which, I will repeat, is being consistently violated. Knowledge in conditions, as with all the principles of war, is essential to its application, but, whilst in the case of several of these principles it is difficult to arrive at the value of war conditions, the conditions I have mentioned are not difficult to grasp or to create, and on how far we are able to create them during peace-time will depend our endurance during war.

### 13. The Expenditure and Maintenance of Endurance

In war moral force is expended in the form of moral friction or explosion, mainly caused by physical danger and loss, and mental misunderstanding of the conditions which surround it. It is maintained by removing danger, by establishing comfort, and by the solidarity of order and organization.

In place of examining these minor though all-important conditions, I will now turn to endurance in a higher form.

The will of the general-in-chief and the will of his men must endure—that is, this dual will must continue in the same state, and in war local conditions are continually weakening this state and threatening to submerge it.

To the commander endurance consists, therefore, in power of overcoming conditions by foresight, courage, and skill. These qualities cannot be cultivated at a moment's notice, and the worst place to seek their cultivation is on the battlefield itself. To enter a battle with a failing heart and an empty head is far worse than bringing a gun into action with a lame team and an empty limber.

The commander must, therefore, be a mental athlete, for his dumb-bells, clubs, and bars are the elements of war, and his exercises the application of the principles of war to the conditions of innumerable problems. In the past this has seldom been done, and many noted generals have spent years in an army, and have had statues erected to their memory, who never touched a dumb-bell or even carried out a mental goose-step.

In an army endurance is intimately connected with numbers, and, paradoxical as it may seem, the greater the size of an army the more difficult is it to maintain its moral solidarity; for, as size reduces speed of movement, so does size reduce speed of thought and increases the area and speed of fear. The reason for this is a simple one. One man has one mind; two men have three minds, each his own and a crowd or group "mind" shared between them; and the larger the crowd the more difficult is it to control the crowd rationally, and the less it is controlled the more susceptible to instinct does it become. If a task which normally requires a thousand men can be carried out by one man, then this one man, morally, will possess a much higher endurance than any single man out of the thousand.

Physically, endurance has little to do with numbers, for the greatest encumbrance on the battlefield is man himself. One invulnerable man is worth a thousand vulnerable ones, and, though complete invulnerability is unobtainable, the principle of endurance, in its broadest sense, should aim at rendering *moral* as invulnerable as possible—that is, the securing of it against the bombardment of the enemy's initiative so that moral force may endure as the mainspring of offensive action.

As the principle of endurance has as its primary purpose the security of the minds of men by shielding their *moral* against the shock of battle, inversely the principle of demoralization, or of surprise, aims at the destruction of this *moral*. First, in the moral attack against the spirit of the enemy's nation; secondly, against the plan of its commander-in-chief, and thirdly, against the *moral* of the soldiers under his command.

Hitherto the third, the least important of these objectives, has been considered by the majority of soldiers as the main objective of this great principle, and in the last great war the result was that the attacks on the remaining two, being overlooked during days of peace, were only slowly developed during the days of stress which followed the outbreak of hostilities.

Since wars are no longer duels between armies, but struggles between nations, the moral attack on the enemy's national spirit is becoming more and more the first and decisive object of a war; and, whatever may be considered legitimate warfare to-day, it is all but a certainty that the energies of the next great war will mainly be directed against this objective, and relentlessly waged by every means at the disposal of the belligerents. This being so, let us as a nation be on our guard lest we become demoralized even before war be declared, for on our national endurance will depend the future success or failure of our arms.

To discredit the policy of our enemy's government is our second moral objective; this is accomplished not only by raising internal discord, but by a persuasive propaganda amongst our enemy's allies, active and neutral. By forcing these allies to bring a disruptive influence to bear, we undermine our adversary's political power, we force him to modify his policy, and, through these modifications, we cause a disruption in the plans of his general staff, and thereby undercut the moral stability of his troops.

The controller of fear is *moral*. In the past *moral* has been attacked by gunpowder; in the future the indirect and unseen weapons of insidious propaganda will, I think, play a far more dangerous part.

The physical strength of an army lies in its organization, controlled by its brain. Paralyse this brain and the body ceases to operate. Paralysis may be creeping or it may be sudden; the first constitutes the moral attack, the second the moral assault; both of which are resisted by putting into force the principle of endurance.

### 14. The Principle of Security

I now come to the third of the principles of resistance, namely security, which is the base of offensive action. "What is the object of defence?" asks Clausewitz, and he answers: "*To preserve.*"[1] To preserve what? The endurance of offensive action expressed in the determination to win, which presupposes movement. According to Jomini, "He who awaits the attack is everywhere anticipated."[2] This is true unless the waiting side is so secured that conditions are against the attack succeeding. If it is not so secured, then to await the attack is a violation of the principle of security. From this it will be seen that it is difficult to determine where the principle of security begins and ends; but, though this is more clearly apparent in the case of this principle, this difficulty exists with all the remaining principles since one merges into the other, and the complete nine into the law of economy of force.

When the mind wishes to stabilize itself it takes up what may

[1] *On War*, vol. ii., p. 134. [2] *Art of War*, p. 73.

be called a "protective attitude," and to give tangible expression to this stability it first makes a demand on moral resistance, and, secondly, on physical resistance. The security resulting from these two forms the base of all offensive action, and the reason is that normal man is influenced in a far higher degree by his instinct of self-preservation when confronted by danger than by a desire to assert himself. During peace-time this moral condition is invariably overlooked, for, as danger is absent, the instinct of self-preservation remains dormant, and the soldier becomes bellicose in the extreme. He demands all kinds of offensive weapons, plans every manner of offensive tactics, and is for ever pommelling his imaginary enemies *because they are imaginary*. Once replace these images by living men armed with lethal weapons, and instantly the soldier performs a mental somersault and seeks to secure his life the moment it is threatened. We must remember this, for otherwise the whole of our peace-training will be based on faulty premises. It is one of the greatest of errors to believe that teaching men how to protect their lives and to set a value on security will induce them to become cowards on the battlefield. On the battlefield men are always cowards, or, if this word appears too strong, then prudent people. The man who does not mind being shot at is a dangerous lunatic; also the man who does not know how to protect his life is going to "let his leader down" by getting shot at the very moment the leader requires his services most—that is, when he is in the greatest danger. If a soldier thinks the instinct of self-preservation can be abolished, then he should resign his commission, for there is no place for him in an army—not even in a base store. In place, as I have already suggested, we should utilize this instinct by turning it into an alarm-bell which will awaken protective reflex action which, when fear seeks expression, will unconsciously and instantaneously suggest to the soldier an act which will lessen the danger without impeding his progress.

We must not confound these conditions of security with *moral*. *Moral* is the force which, by balancing fear, allows determination to impel the soldier forward. *Moral* includes patriotism, esprit de corps, comradeship, confidence, loyalty, etc., all of which are acquired qualities and virtues. These are being sapped and undercut by fear. The means are, consequently, those actions and physical things which shield *moral* from these attacks. In a highly-trained soldier the most important of these actions is the offensive itself, and why? Because the most certain security is attained by defeating the enemy—that is, by removing the cause which, as its effect, awakens fear. The real battlefield is inside the skull and not outside it, and as the brain is the best protected of all the organs, being completely armoured by the skull, so must *moral* be completely armoured by the application of the principle of security to the conditions which surround the soldier. A sudden blow on the head will stun a man ; a sudden blow to *moral* will frequently stun the individual soldier, and sometimes stun even an entire army. It is against such blows, small and great, that men must be secured ; physical blows are but the left-hand punches which culminate in this right-hand blow to the jaw, and the jaw may not necessarily be a military one.

In all pursuits mind is the directing force. To the actual combatants this directing force expresses itself in determination. As *moral* is used up self-preservation takes charge, consequently unless *moral* is economically expended there is always a chance that we shall fail to gain our object.

The objective we have set ourselves is our goal ; determination is the propellant we use to gain it. The maintenance of *moral* is not in itself an objective, but a means to gain our object ; we have got to expend moral force, and in modern wars, in which whole nations are concerned, it must not be forgotten that all military action is but a means of securing the prosperous existence of the nations at war. I intend, therefore, before dealing with the purely military aspect of the principle of security, to hark back to chapter iv., and examine how this principle can be applied in order to shield the gaining of the ethical, national, and economic objects of war.

### 15. Ethical Security

In chapter iv. I dealt at some length with the non-military objects of war. As regards the ethical object, I pointed out that the winning of it formed the true foundations of peace, consequently it is worth securing.

During the Great War a battle of propaganda was waged by all belligerents, though at its beginning few were prepared to wage it. Our object was to prove that the Germans were " dirty dogs," and that it was they who had started the war. I do not suggest that our contentions were wrong, but I cannot help feeling that when the Germans retaliated the means we employed to protect our national character were not of the best. In place of maintaining our reputation for fair play we hired a pack of journalists to defend us. These people, who had spent their lives in raking filth out of the law courts, went to mud with the alacrity of eels, and, though they undoubtedly succeeded in blackening the German nation, we ourselves became somewhat piebald in these gutter attacks.

The point I wish to accentuate is that propaganda is not only a powerful weapon, but that it becomes a two-edged one if held by an unclean hand. The Germans had committed sufficient crimes for us to pillory them publicly, but to accuse them of nailing babies to barn doors and extracting margarine from dead soldiers was to smother ourselves in ridicule. By such means a nation cannot secure its character against attack; it may injure its enemy, but in doing so it injures itself. A liar, be it well remembered, is a moral suicide.

### 16. National Security

National insecurity is one of the fundamental causes of war, especially if the nation concerned is militarily powerful. All nations are impelled by the instinct of national preservation to seek secure frontiers, and, if secure frontiers cannot be gained by peaceful methods, powerful nations will seek to secure them by war. A strong frontier is nothing else than a natural fortress, which, when garrisoned, secures the nation against attack. The object is the security of the nation, consequently, as I have already pointed out, the breaking down of the national will is the surest means of forcing the fortress to capitulate.

Up to quite recently nations could only be attacked on land, or on the sea if they were not self-supporting, but to-day they can be attacked from the air. This possibility has introduced a problem of security which must revolutionize the whole military outlook.

Direct protection against aerial attack is purely a military problem, namely command of the air, so I will not consider it here. Indirect protection is a civil problem; in other words, the civilian population must protect itself by so organizing itself that its *moral* can withstand a series of terrific nerve shocks.

The main weakness in the nervous system of great nations is to be sought in the concentration of vast numbers of people in towns, the dependence of these people on regular traffic, and the rapidity with which a disaster may become contagious by use of the post-office, telegraphs, and telephones.

To apply the principle of security in existing conditions is most difficult, because they are such as render the contagion of panic almost electric in its swiftness. It would consequently appear that the solution of this problem lies in being prepared at a moment's notice to isolate panic by switching off the whole of the intricate system of communications which brings every part of a country in time within a few seconds of each other; or, to put it still more plainly, to paralyse temporarily the country or district attacked, and, under cover of the inaction resulting, to establish order, and, once it is established, to follow this paralytic stroke up with a flood of reassuring messages. What is here suggested is, not the application of a principle to a series of most difficult conditions, but the instantaneous creation of a condition in which the principle can most readily and rapidly express itself. On the battlefield this is seldom possible, but I see no reason why, amongst the peaceful population beyond the battlefield, such a method should not work successfully.

### 17. Economic Security

To ourselves, a non-self-supporting country, the importance of economic security is too obvious to require accentuation. In war we have got to secure ourselves against loss of food supplies, loss of markets, and loss of internal resources. Attacks on these may be either direct or indirect; in the first case, such as attacks on our overseas trade by surface craft, submarines, and aircraft; in the second, by extortionate prices asked for war necessities not provided by the country itself, and an unscientific use of all resources by the defence forces.

I have already dealt—in chapter iv.—with war economics, and, whether the factors I have quoted are correct or not, this in no way vitiates the importance of the higher command of an army realizing that economy is essential in war. If gold is the sinews of war, and gold, as money, is only " potted " man-power, or work, then every coin badly spent—that is, uneconomically spent—is a sinew injured. To prepare soldiers to exercise economy in war it is essential that they should be allowed financial responsibility during peace-time. Without such responsibility, though the necessity of economy may be appreciated, the means of effecting it will not be understood, and in war they cannot be learnt. To-day we are not only paying for the cost of the war, but for the parsimony which preceded the war. We are paying for our previous lack of economic " backsight," insight, and foresight, and our ignorance of how to secure ourselves against self-inflicted economic injury.

### 18. Military Security

I have dealt with these three non-military forms of security because, throughout this book, I wish to impress upon the student the importance of realizing that war is a national and not merely a military activity. The entire military power of a nation is based on civil power, and never more so than to-day, when in war nations are *nations in defence*. I will now turn to the military aspect of the principle I am examining.

Napoleon once said—and his words are full of truths—that:

> You should make a start from such a powerful defensive order that the enemy will not dare to attack you. . . . The whole art of war consists in a well-reasoned and extremely circumspect defensive, followed by rapid and audacious attack.

The soldier should learn this saying by heart, and do more than remember it, for he must *understand* what it means. It means that the foundation of success is strength, and that offensive action is based on defensive power. Every action or movement forms the base of the next action or movement, consequently every action is related to the last, and must be considered with reference to the next. Thus alone can the object be maintained. There in front of us is the ultimate goal, and each move in the game is to gain it. Every action secures the next action. It is because of this interplay that the principle of security never ceases to operate; it is always operating, but unless we can control it its activities may defeat us.

The object of battle being to destroy the enemy's fighting strength, that side which can best secure itself against the blows of its antagonist will stand the best chance of winning, for by saving its men and weapons it will augment its offensive power. Security is, therefore, a shield and not a lethal weapon, and to look upon it as a weapon is to turn war upside-down. Consequently the defensive is *not* the stronger form of war, but merely a prelude to the accomplishment of the military object of war—the destruction of the enemy's strength by means of offensive action augmented by defensive measures. What is the stronger form of war is a well-secured offensive operation. I mention this here because, in the minds of some, defensive warfare is still held as the stronger form, the reason being the terrible losses all parties recently sustained in attacks on trenches. These people are obsessed by the idea that the whole art of war consists in constructing a Chinese wall of fire-power; of letting the enemy attack it, and commit suicide by doing so. The final actions of the Great War should have dispelled this illusion.

As the offensive is essential to the successful attainment of the object, it stands to reason that security without reference to offensive action is no security at all, but merely delayed suicide. Every man needlessly employed in defensive work is a weapon-wielder less for offensive operations. In order to avoid an excessive use of men for purely protective duties recourse is had to guards and outposts, the strength of which depends on the condition of time. The time it will take an enemy to cover a certain distance, or the time it will take his opponent to frustrate him doing so. Security, therefore, may be frequently considered as simply a means of gaining time at the expense of the enemy.

As danger and the fear of danger are the chief moral obstacles of the battlefield, it follows that the imbuing of troops with a sense of security is one of the chief duties of a commander; for, if weapons be of equal power, battles are won by a superiority of nerve rather than by a superiority of numbers. This sense of security, though it may be supplemented by earth-works or mechanical contrivances, is chiefly based on the feeling of moral ascendence due to fighting efficiency and confidence in command. Thus, a man who is a skilled marksman will experience a greater sense of security when lying in the open than an indifferent rifleman in a trench.

Given the skilled soldier, the moral ascendency resulting from his efficiency will rapidly evaporate unless it be skilfully directed and employed. As in all undertakings—civil or military, ultimately we come back to the impulse of the moment, to the brains which control impulse and to each individual nerve which runs through the military body. To give skilled troops to an unskilled leader is tantamount to throwing snow on hot bricks. Skill in command is, therefore, the foundation of security, for a clumsy craftsman will soon take the edge off his tools.

## 19. Strategical Security

The basis of strategical security is the soundness of the plan of action, the logistical distribution of the troops, the maintenance and correct location of the reserves, and the protection of the lines of communication. Other factors which influence strategical security are infrequency of change of objective, or direction, and the absence of unnecessary movement.

Strategical security is also attained by placing an army in a good position to hit at the communications and headquarters of the enemy whilst protecting its own—by so distributing a force that it may live at ease and fight efficiently. Though movement, actual or potential, is the soul of strategy, the placing of forces in the area of operations so that their very position threatens the enemy's initiative is the spirit which should imbue all generalship, for on it rests the security and offensive power of an operation, a campaign, or of a war.

The relation between strategy and tactics is one largely governed by the principle of security and the principle of distribution of force. Strategical distributions and operations aim at securing tactical action. They do not merely protect it, but they enable it to take place, either directly through movement, or through co-operation or combination of forces. In the battle itself security is effected by tactical action, but before the battle the strategical distribution is the " defensive order " as employed by Napoleon, which I examined in chapter viii.

### 20. Tactical Security

Grand tactical security may be defined as " the choosing of a vulnerable target or the refusal to offer one." Here the factors are mainly those of time and space. The rapid employment of weapons at the decisive point, whether for attack or defence; the general organization of battle—the penetration of a front, the envelopment of a flank, the endurance of the fight, whether by retirement or pursuit—those and many other actions build up that general security which cements the units of an army into one co-operative whole.

Minor tactical security embraces the entire gamut of a soldier's actions—his *moral* and efficiency, the quickness and audacity of his leader, the judgment and determination of his commander, and the confidence of his comrades. On the battlefield itself security will depend on seeing and not being seen, on hitting and not being hit, on moving and not being moved. The first embraces surprise, observation, and cover from view; the second the use of weapons, ground, and armour; the third mobility and protected movement. To move quickly is to reduce the chance of being hit. To suit formations to the conditions of fire and ground is simultaneously to increase hitting power and to reduce the vulnerability of the target.

In all tactical action surprise offers the most effective means of securing an attack or of breaking down security. In all circumstances it must be applied and guarded against. As a surprisal is an operation which seldom permits the party surprised time wherein to carry out a deliberate counter-move, all troops should be trained to execute certain counter-actions automatically on being surprised. Though these may not always be the most suitable in the circumstances, it must be realized that the power of surprise lies in stunning the reason. Men have no time to think: Shall we do this, or shall we do that? Leadership on these occasions is frequently reduced to zero, consequently to prevent chaos intervening—and it is loss of order which is the true enemy —soldiers should be trained to carry out collectively and spontaneously a definite move to meet a certain type of surprise. Thus—to take a few examples—what action will best prevent loss of order, and consequently of control (not necessarily loss of life), in the following circumstances: an aircraft attack against a marching column; a tank attack against a deployed line; a sudden attack in front or in rear? The point to note is that immediate counteraction is not necessarily offensive action, that it should aim not so much at protecting the man as protecting the organization. When a salvo of shells falls near a company in close order it does not scatter into human dust, but into sections in artillery formation. If necessary, directly the danger has passed the whole company can, in a minute, be re-formed in its original close order. This is a good example of what I mean by automatic counteraction, or the security of organization against a sudden surprise. To-day we have many new weapons against which there must exist some counteraction, though these various means of securing local command may in no way be offensive in nature. These means must be thought out and practised.

### 21. New Problems of Security

I have just made mention of new weapons, and I will end this chapter by considering their influence on existing methods of security. The changes which these weapons (especially gas, the aeroplane, and the tank) are daily creating are radical. I cannot examine them in detail in the space at my disposal, but I can take two or three examples, and by means of these show how completely our former ideas are being changed.

I will first examine the elastic square. To protect itself an army throws out an advanced, a rear, and two flank guards, sufficiently far from the main body so that, if one or more of these guards is attacked, the main body will have time to deploy. To-day the aeroplane can "hop over" these guards, and in a few minutes attack the main body, which, to secure itself, will have to add a fifth guard to its existing four—a sky, or air, guard. Such a guard must consist of aircraft which, offensively, are immobilized whilst employed on this protective work. If the column is a mechanical one it can be armoured, and if tracks, in place of wheels, are used it can move across country, and so reduce the size of the target it offers. In the past indirect protection against the bullet was sought by extensions, and direct protection by cover by ground. An advanced scout signalled the approach of the enemy, and the troops extended and took cover. In future, though the means will have altered, the application of the principle of security will be identical. An aeroplane ten miles away will signal hostile aircraft; in a minute or two the mechanical column will take up anti-aircraft extensions, and, in place of ground, will use armour.

I will take another case, which may be represented by the letter T. The vertical stroke represents a column halted for the night, and the horizontal stroke represents its outposts. Hitherto the distance between the outpost line and the main body has been calculated on the time factor—the resistance required to gain sufficient time to enable the column to deploy. Against an infantry attack the outposts may resist for several hours; against a tank attack they will be overrun in a few minutes; and perhaps a quarter of an hour later the main body will be attacked, when it is in no way prepared to meet an attack. Should this column be a mechanical column there will be no necessity to deploy, for it will rest deployed. The application of the principle of security is exactly the same, but the conditions in which it is applied have changed.

Here is another example. Six good roads exist in a certain area, and these are to be used to concentrate three army corps at a definite locality at a definite hour. The enemy, by means of tanks, soaks a mile or two of three of these roads (at places where they run through defiles) with vesicant and lachrymatory chemicals. The result is that three divisions are delayed for twenty-four hours. A mechanical column can move off the road, or, if its machines are gas-proof, it can move straight ahead. The application of the principle of security is the same—namely, avoiding the danger—the only difference is that one type of column can avoid it more speedily than the other.

These three examples will be sufficient to illustrate the type of changes in security which are now taking place—not changes in principle, but in the conditions of war. Whilst ten years ago security was in nature mainly lineal, to-day it is no longer so, and the principle of security has to be applied to entire areas as well as to battle-fronts, consequently to entire nations as well as to armed forces. Unless we understand these changes we cannot apply this principle of security, and unless we can apply this principle we cannot apply the remaining eight.

## CHAPTER XV

### THE APPLICATION OF THE SCIENCE OF WAR

The heights by great men reached and kept
Were not attained by sudden flight,
But they, while their companions slept,
Were toiling upward in the night.
—LONGFELLOW.

### 1. A BRIEF SUMMARY OF THE METHOD

I HAVE now outlined the foundations of my system, which, with all its faults—and it must possess many—is an attempt to establish the theory and practice of war on a scientific footing by applying the method of science to the study of war. I do not claim to have discovered any talisman which will protect the soldier against defeat, or charm him to victory; but what I hope I have done is to convince him that war can be reduced to a science, and must be so reduced before, as an art, its forces can be correctly expended. Further, I feel that it is through system that study becomes interesting, and, because of the lack of system, military history, though read, has been of so little value to the soldier, for many have profited from it no more than the old Mandarin general in Mr. Flecker's " Golden Journey to Samarkand ":

Who never left his palace gates before,
But hath grown blind reading great books on war.

Had he studied war on a system which would have enabled him to have discovered why certain actions failed and why others succeeded, his eyes might have been opened. I will therefore now summarize very briefly a few of the salient points in my system, and then show how it can be applied to the study of military history, or to the development of a plan of campaign or battle, or to the solution of any tactical problem or exercise.

The causes of a war enable us to obtain an insight into its nature—that is, the type of war fought or to be fought—and on this nature depends the political object of the war. This object should direct the policy of the government, which should be put into force by the plan of the general-in-chief. In its turn, the plan is determined by the military object, the gaining of which demands expenditure of force, and expenditure depends on the conditions which surround and influence the instrument of war, whether it consists of all three fighting Services or only one Service.

Thus we are reduced to three military requirements of the first importance:

(i.) Knowledge of the powers and limitations of the instrument.
(ii.) Knowledge of the powers and influences of conditions.
(iii.) Knowledge of how to expend force profitably.

The instrument includes three forces—mental, moral, and physical force—which must be organized before they can be profitably expended. Organization demands a definite structure and maintenance, and when these two are in harmony organization can be controlled.

Conditions influence the three forces of the instrument, therefore the conditions of war may be divided into three categories, whether these conditions be material or human.

The problem now resolves itself into discovering:

(i.) The elements of the forces of the instrument.
(ii.) The influence of the conditions on these elements.
(iii.) The law which governs changes of force in the elements as conditions influence them.

The forces of the instrument I have reduced to nine elements.

(i.) Mental elements: reason, imagination, and will.
(ii.) Moral elements: fear, *moral*, and courage.
(iii.) Physical elements: weapons, protection, and movement.

The influence of conditions are that they can assist, resist, and transform the force of each element, and through them the nature of the instrument.

To discover the law which governs the changes of force I turned to physical science, for if the laws of uniformity and causation govern all forces in the universe, they must also govern the expenditure, or changes, of forces in war. I learnt that all changes of force were expressed in motion, and that all motions were the resultant of the pressure and resistance exerted by one or more forces on another, and I called the law which governs these changes the law of economy of force.

As this law governs all changes in force, the next question is how to apply it.

We know that all forces are continually in tension—that is to say that they are ceaselessly pressing and resisting one another. We know also that when the conditions in which tension takes place are the same this tension does not vary. Consequently, if we know what this tension is and how conditions influence it, when certain known conditions occur we can expend our force economically, that is as it would be expended were our wills replaced by the law of economy of force.

As in war the forces of the military instrument find their tension in three spheres, and as each can be reduced to three elements, we obtain nine general expressions of the law of economy of force, which I have called the principles of war. These principles are abstract generalizations of the tensions within the elements caused by the varying influences of the conditions of war.

In the mental sphere we direct, concentrate, and distribute force in idea, and base our actions on these ideas ; this gives us the general outline of our plan.

In the moral sphere we adjust this outline according to a more detailed examination of the elements of this sphere as influenced by the conditions of war, and the principles of direction, concentration, and distribution change into those of determination, surprise (maximum power to exert moral pressure), and endurance (maximum power to resist moral pressure).

In the physical sphere we carry this adjustment of our plan to its conclusion by examining in detail the influence of conditions on the physical elements of this sphere. Determination now evolves into mobility, and surprise and endurance into offensive action and security.

Thus does the law of economy of force, in the form of the principles of war, ceaselessly operate through all the spheres of force, whether we apply this law or not. If we fail to do so, by attributing an erroneous cause to an effect, or vice versa, then our plan will fail to synchronize with this law, and punishment will be meted out to us in *exact* proportion to our errors.

As, generally speaking, the fewer the parts of any machine the simpler becomes its working, it is, as I have attempted to show, an assistance to rapidity of thought to arrange the nine principles of war into three groups, not according to the nature of the spheres of force, but according to the functions of force in each sphere.

Thus the interplay between the faculties of reason and imagination controls the will by directing it. The interplay between the sentiments of fear and *moral* controls courage by determining its value. And the interplay between the physical means—weapons and protection—controls movement by regulating its mobility.

We thus obtain a compound idea of control by uniting the principles of direction, determination, and mobility. Similarly, by uniting those of concentration, surprise, and offensive action, do we obtain a compound idea of pressure; and by uniting those of distribution, endurance, and security, a compound idea of resistance.

Finally, my whole system can be concentrated into seven words:

Cause—Object {Elements, Principles, Conditions} Objective—Result

The cause may be either the cause of a war or of an order received, and the result is the terms of peace or the effect of our actions in carrying out the order. The object is our intention, and the objective is gained when our intention is fulfilled. The elements are the forces at our disposal, and the conditions all forces which influence them; and the principles are our guides, and the law of economy of force is our master.

## 2. The Study of Military History

The study of history of any kind is always difficult, not only because the human factor is so pronounced, but because the atmosphere of past events is not the atmosphere we breathe to-day. Reliability of evidence is the first requisite, the second being the reality of conditions in which the event described took place.

In military history these difficulties are accentuated by the fact that evidence is based largely on the reports of eye-witnesses, which at the time cannot be subjected to criticism, and that the atmosphere of the battlefield is so tremulous with excitement that those who have breathed it are frequently at a loss to reproduce it even in memory after the battle is ended, and as time lapses its influence is rapidly forgotten. If this were not the case, we should not so often see during peace-training the amazing determination which is displayed and the total scorn of danger.

It is in peace-time that such terms as the following are invented: "to the last man and last round"; "dying in the last ditch"; "holding a position at all costs"; "to die at your post," etc.,

etc. But in war most of us sympathize with the boy in *King Henry V* who exclaimed: "Would I were in an ale-house in London! I would give all my fame for a pot of ale, and safety;" for "The groan, the roll in dust, the all-white eye turned back within its socket" is a reality we are unaccustomed to in peace-time.

Before the outbreak of the war in 1914 I happened to be a student at the Camberley Staff College. At the time I had evolved part of my present system, and was appalled by the way I was expected to learn rather than study military history. It appeared to me to be done backwards. So to speak, we got into Mr. Wells's "Time Machine," and, carrying with us a big chunk of Camberley atmosphere, we set out, not for the Elysian fields, but straight for the Shenandoah Valley, never dreaming that a far more important war, namely the next war, the only one we could take part in, was ever going to be fought. To the Shenandoah Valley we went without really going there, and we carried with us an immense number of brain-sacks and a huge shovel. And what did we do when we got there? When we got to that place, to which in reality we never got to, because of the Camberley atmosphere, we shovelled facts and fictions into those sacks, pell-mell, to bursting-point, and then we came home and played golf! So many facts did I collect on the Valley Campaign that I believe, had I been asked the weight in kippers Stonewall Jackson ate for breakfast on the seventeenth Thursday of the year 1862, I should have answered off-hand: Five-sixteenths of a pound; and would have been right to within a quarter of an ounce.

This may be considered to be harsh criticism, but it is not intended to be solely destructive, for I have attempted to replace the system of 1914 by what I believe to be a better system, and I hope that twelve years hence my system will be as heavily attacked as I have attacked the one I suffered under; because it will show that progress has been made and the faults in my system have been discovered.

Who invented this extraordinary method of absorbing ink visually I do not know, for to find a parallel to it, would demand a return to the study of theology in the Middle Ages.

I have not related this personal experience as a digression, since the system of 1914, if not so vigorous, is still the system of to-day. The first fact to note is that the study of history possesses only one true value, the discovery of what may prove useful in the future. The object of the study of history is to prepare us for the next war, consequently all the ephemeral details of 1862, etc., should be passed over lightly, and attention concentrated on what is of permanent value in war. What is required is the "why" and "how" of success and failure in a series of campaigns, and not the microscopic knowledge of any one campaign.

To return to the two initial difficulties. The best evidence is not local evidence, but distant evidence, and the evidence supplied by military writers who have had experience of war. Xenophon's history of Cyrus and Arrian's of Alexander, though not necessarily true in all respects, are models in reality. We not only listen to the historian, but we see the hero. The Cyrus of Xenophon is almost a fictitious character, nevertheless he is real, for such men do exist, and Xenophon was one of them. If now we turn, for example, to Dr. Conan Doyle's *History of the Great War*, which runs into several volumes, I cannot imagine any soldier discovering one item of value in it.

When, however, we turn to the second difficulty, the atmosphere of a war, Conan Doyle's work might help one to appreciate the astonishing superficiality of knowledge in an educated civilian of imagination during the years 1914–18. To breathe the atmosphere of war we must read books of the period, books written during the war or immediately after it, but with circumspection. For instance, Sir Philip Gibbs's *Realities of War* gives one a wonderful description of the sentiments of a maiden aunt in Upper Tooting shell-shocked by the *Daily Mail*, but it has nothing to do with what the soldier felt in France, for, though its writer was in France, he was still breathing the air of Tooting Bec.

Official histories convey no reality and no atmosphere, but only facts. To obtain atmosphere the memoirs of some gay and human soul, such as Samuel Pepys, should be first read, and then, when psychological insight into the period has been gained, the leading historians should be studied methodically.

### 3. The Application of Method to Military History

For reasons mentioned in the Preface of this book, I do not intend to examine an actual campaign in the light of my system; instead, I will briefly outline how I should proceed in this examination.

To understand the nature of a war, and it is its nature which determines its procedure, a clear grasp of the causes of the war is essential, and especially so in modern times. These causes are difficult to discover, since military historians are so apt to be prejudiced in favour of one side or the other, and political historians generally confuse pretexts with causes, and general historians, knowing so little about war, normally consider its outbreak as they would a cataclysm—an earthquake or a flood.

In small or local wars the difficulty of distinguishing causes is insignificant when compared to the discovery of the causes of a great war, for, whilst the origins of wars in the second and third degree are generally traceable to a clash of opinions, those of great wars are wrapped up in biological and psychological influences.

In examining the causes of a great war it is wise, I think, to go back to the last great war which preceded it, and to examine in detail the peace treaty which concluded it, and from its military, economic, and ethical aspects. It is worse than useless to begin our search in the period immediately precedent to the outbreak of war, since this period, politically, is a mass of lies, and, if we do begin by studying it, unless we are very careful, we shall be misled by the clever and calculated attempts of pots and kettles calling each other black.

Once we have settled on the causes, we next discover the object of the contending parties ; what is their political intention ? On one side there must be a definite aim, if not on both. To discover this intention it is not necessary to wade through many books, but to examine the mentality of the most influential statesmen and soldiers and the general outlook of nations.

If we study history with our eyes open, we soon discover how restricted are the influences of the masses, for, however democratic, socialistic, or communistic they may be, they are inarticulate. In place we find events revolving round a few leading personalities, more frequently than not philosophers, poets, men of science, etc., rather than politicians and soldiers. For instance, Hegel, Byron, Darwin, and Nietzsche had far more influence in fashioning the "mentality" of modern Europe than all the politicians and soldiers of the last century. It is men who, like William Blake say, " I must create a System, or be enslav'd by another Man's," which fashion the inner intention of war, which finds its tangible form in the political object of war itself.

We now arrive at the military phase of our study ; the military objects, however unscientific they may be, are discovered in the respective plans of campaign, the values of which mainly depend on the ability and character of the opposing generals-in-chief ; for their character should stamp themselves, not only on the plan, but on the armies they lead. If this is not the case, then we may be certain that they lacked personality, and were only figureheads.

The next problem is to evalue the respective instruments of war—their organization, nature, and potential activities ; for from these will their strategy and tactics be developed. Our sieve has nine compartments, and, in place of shovelling facts into brain-sacks, we should throw them up against this imaginary series of grids. This will enable us to sort out facts according to their values. Thus we start with all the movements which take place, whether strategical, tactical, or administrative. The next two grids give us their offensive and protective values, the next three their moral values, and the last three their mental values.

In examining movement we should first enquire into the physical nature of the opposing sides; there may be no intrinsic difference, but there may be an artificial one, such as the load carried. Our study may perhaps tell us that the load was too great, that equipment was thrown away, that marches were short, or that men fell out in numbers. These are important points, for man's physical strength does not vary much, consequently here we discover not only lessons of the past, but also for the future. Muscular energy is economized by mechanical means of movement. What has the campaign to tell us about these? The railways, the roads, the rivers, the canals, the sea, and the air; what were the main influences of these means on the campaign; how far did they assist the commander, and how far did they complicate his work? Deficiency of means of movement should also be studied under this heading.

The next pile of facts we should examine are weapons. Probably both sides are similarly armed, but possibly the employment of weapons differs. If so, what are the value of the differences, as well as the value of the weapons themselves? If a new weapon was introduced, what really was its value? Did it simply gain a fictitious reputation because one side had it and the other had not? What were its influences on existing weapons, on tactics, and on *moral*? The normal historian will tell us little about all these things, but by reading between the lines we shall discover a point here and another there, and by degrees accumulate valuable facts.

In the physical sphere our last question is protection. On what theory of protection are the two sides working? Is it direct or indirect, static or mobile? What are their various means of carrying out their theories concerning extensions, smoke screens, camouflages, trenches, obstacles, fire-power, armour, etc., etc.? What are their respective values in varying conditions? What appear to be their weak points and their strong points?

From all these considerations we obtain a tactical structure, and then from the physical we turn to the moral sphere of war, which animates it and maintains its force.

Leadership, based on an encouraged will, is the next problem to examine. What are the theories of leadership? They differ in most armies ; in some they are autocratic ; in others democratic. What is leadership based on ? Is it fear, or affection, or the intelligent use of means ; is it all three, or which one in particular, and how do the men respond to each type ?

This leads to the nature of the *moral* of each side. What is its nature ? This depends first on national characteristics, and only secondly on military training. The nature of national *moral* differs, and sometimes considerably. Not only do we want to know the nature of the eventual theatre of war, but the nature of the beasts we are going to meet in it. Is not this what every hunter does ; he differentiates between the instincts and individual characteristics of the various animals he intends to hunt, and varies his actions accordingly. If we are going to fight Turks, we want to know what is meant by *a* Turk, if Germans, then equally do we want to know what is meant by *a* German. The one thing that we do not want to do is to mistake a rhinoceros for a gazelle, because those who do so seldom survive to make use of their experience.

From national *moral* we next turn to military *moral*, and try to discover what is the doctrine of discipline—that is, the mental and physical machinery used to convert the man into a soldier. Discipline should accentuate the virile national characteristics, and tone down the effeminate ones. Is it based on fear or comradeship ? Does it aim at cultivating initiative, or of subordinating the individual will, or in stamping it out ? When we turn to our campaign we shall see how the respective doctrines stood the test of war.

We have now discovered the factors which animate the armies, and so can turn to the mental sphere of force which controls the instrument.

Here the main problems centre round the general-in-chief. Is he a free agent, or is his will shackled by political control ? What are his reasons for his various moves, and do these reasons display originality and imagination ?

We have now completed the first phase of our study, and the next consists in an examination of the conditions the war was fought under. How far did each side appreciate the nature of conditions before the campaign started, and whilst the war was in progress ? Unless we grasp this we shall frequently be misled, and we shall seldom grasp it unless we carefully analyse what the conditions are, irrespectively of the actions they eventually influenced ; for the more we realize the true nature of these conditions the better are we able to appreciate the value of the actions fought.

The conditions of war having been analysed, correlated, and

surrounded by an atmosphere of reality—as far as in our imagination we can recreate the atmosphere which existed at the time the campaign was fought—we should next shovel our nine little elementary heaps into one heap and place it in position on the map. Then, in turn, we should play a game of Jekyll and Hyde. For half an hour we are Napoleon, for another half-hour Blücher. To play this game properly we must see and think, as far as we are able, as these generals saw and thought, and we cannot do so unless we understand their personalities.

Now as to the application of the principles of war. At first it may be thought that, as in many campaigns the contending commanders had little or no knowledge of the principles of war, if we are to play the parts of Jekyll and Hyde, how are we going to learn to apply them?

The fact is that, besides fulfilling this dual rôle, we have got to play a third part, the part of a disinterested critic and judge. We must remember that, though at the time in question the value of the principles of war may have been unknown to the opposing commanders, they, as truths, nevertheless existed, and that their unconscious application or violation resulted in success and failure, even if reasons for success or failure were not apparent at the time.

It is by discovering these reasons that we add to creative thought. We accomplish this by constantly asking ourselves the questions: What was the object of that move? What was the concentration of force attempted? What was its distribution and its direction? Thus at the battle of the Marne, Maunoury attempted to attack von Kluck's right wing. Why did he do this? Would not an attack on the left wing, or a holding attack by frontal pressure, have been more effective? Each must be weighed against the existing conditions—ground, *moral*, position of other troops and communications, etc. Then to each alternative objective we must apply the three physical principles of war; this will give us the outlines of a series of possible strategical and tactical actions. We must then paint in the detail by applying the three moral principles and the three mental principles of war, and so obtain a series of finished pictures or plans. How are we to judge which of these is the best? By turning to the law of economy of force and calculating which will require the least expenditure of force in the gaining.

We must not for a moment imagine that the most economical expenditure means the plan which will require the least number of soldiers or weapons, for this is not necessarily the case. Normally it is by concentrating strength that economy is effected, for a big military balance enables us to expend this strength

economically at the decisive point. The most vulnerable points are those the capture of which will produce the greatest demoralization, first, in the command, and secondly, in the troops. When confronted by a genius like Napoleon, the decisive point of attack is the genius himself. Remove Napoleon from his command during the 1796 campaign and the probabilities are that the Austrians would have won the war.

Though such a removal is seldom possible, the fact to bear in mind is that all operations of war are directed against the enemy's command—*the man* behind the hostile battle-front. Thus ultimately we get back to our starting-point, namely one man.

Unless we can think logically, though we may read the histories of a hundred campaigns and discover thousands of facts, not one may be true. If we desire to derive the greatest benefit from our study of military history, once we have completed the analysis of any campaign, we should project our deductions into the future, and consider their values with reference to the most probable conditions in which the next war will be fought.

### 4. The Application of the Method to Plans and Problems

I will now turn to the question of plans and problems; both can be considered together.

Firstly: We must make certain of our object, or of the purpose of a problem, and whatever we do, we must always refer back to this object or purpose.

Secondly: We must discover the values of our own means and the enemy's by analysing them and deducing the initial power of each element. From these deductions we shall be able to discover the predominant characteristics and limitations of the two instruments.

Thirdly: We must examine the conditions of war and see how they can assist and resist us, and how, on account of their assistance or resistance, the elements in the instruments are transformed.

Fourthly: We must look upon our enemy as a bold and intelligent antagonist who will make the utmost use of his means as influenced by the conditions which will assist him and resist us.

Fifthly: We must apply the principles of war to the enemy's means as influenced by conditions.

Sixthly: We must work out a concise plan, or plans, of action for the enemy.

Seventhly: Bearing in mind the possible moves the enemy may

make, we must apply the principles of war to our own means as influenced by conditions and work out a plan whereby we hope that we can defeat the enemy, and a series of plans whereby we can frustrate the probable moves of the enemy should he gain the initiative.

Eighthly: We must decide on the distribution of our force.

### 5. Maxim for the Ignorant

It is during peace-time that we prepare for war, and, unless our preparation is systematic, unless it is based on some science of war, whether the one I have outlined or some other, for the ignorant—and all are ignorant who do not co-ordinate knowledge—there is one great maxim which throughout the history of war has more often than not proved successful, and this maxim is, "When in doubt, hit out." When the soldier, whether private or general, does not know what to do, he must strike; he must not stand still, for normally it is better to strike and fail than it is to sit still and be thrashed. Therefore I will end this book with a saying of Napoleon's which I have already quoted:

"*The whole art of war consists in a well-reasoned and extremely circumspect defensive, followed by rapid and audacious attack.*"

GUARD

MOVE

HIT

MAR. 69
N. MANCHESTER,
INDIANA

CPSIA information can be obtained
at www.ICGtesting.com
Printed in the USA
LVHW081803100822
725523LV00032B/113